Advanced Concrete Technology

Advanced Concrete Technology

Constituent Materials   ISBN 0 7506 5103 2
Concrete Properties   ISBN 0 7506 5104 0
Processes   ISBN 0 7506 5105 9
Testing and Quality   ISBN 0 7506 5106 7

Advanced Concrete Technology
Constituent Materials

Edited by
John Newman
Department of Civil Engineering
Imperial College
London

Ban Seng Choo
School of the Built Environment
Napier University
Edinburgh

ELSEVIER
BUTTERWORTH
HEINEMANN

AMSTERDAM   BOSTON   HEIDELBERG   LONDON   NEW YORK   OXFORD
PARIS   SAN DIEGO   SAN FRANCISCO   SINGAPORE   SYDNEY   TOKYO

Butterworth-Heinemann
An imprint of Elsevier
Linacre House, Jordan Hill, Oxford OX2 8DP
200 Wheeler Road, Burlington MA 01803

First published 2003

British Library Cataloguing in Publication Data
A catalogue record for this book is available from the British Library

Library of Congress Cataloguing in Publication Data
A catalogue record for this book is available from the Library of Congress

ISBN   0 7506 5103 2

For information on all Butterworth-Heinemann
publications visit our website at www.bh.com

Typeset by Replika Press Pvt Ltd, India
Printed and bound in Great Britain by Biddles Ltd, *www.biddles.co.uk*

(Set of 4) E310

620.136 ADV

Contents

Part 1   Cements

Part 2   Cementitious Additions

3 Cementitious additions   3/3
Robert Lewis, Lindon Sear, Peter Wainwright and Ray Ryle

Part 3   Admixtures

4 Admixtures for concrete, mortar and grout   4/3
John Dransfield

Part 4 Aggregates

5 Geology, aggregates and classification 5/3
Alan Poole and Ian Sims

Preface

The book is based on the syllabus and learning objectives devised by the Institute of Concrete Technology for the Advanced Concrete Technology (ACT) course. The first ACT course was held in 1968 at the Fulmer Grange Training Centre of the Cement and Concrete Association (now the British Cement Association). Following a re-organization of the BCA the course was presented at Imperial College London from 1982 to 1986 and at Nottingham University from 1996 to 2002. With advances in computer-based communications technology the traditional residential course has now been replaced in the UK by a web-based distance learning version to focus more on self-learning rather than teaching and to allow better access for participants outside the UK. This book, as well as being a reference document in its own right, provides the core material for the new ACT course and is divided into four volumes covering the following general areas:

- constituent materials
- properties and performance of concrete
- types of concrete and the associated processes, plant and techniques for its use in construction
- testing and quality control processes.

The aim is to provide readers with an in-depth knowledge of a wide variety of topics within the field of concrete technology at an advanced level. To this end, the chapters are written by acknowledged specialists in their fields.

The book has taken a relatively long time to assemble in view of the many authors so the contents are a snapshot of the world of concrete within this timescale. It is hoped that the book will be revised at regular intervals to reflect changes in materials, techniques and standards.

John Newman
Ban Seng Choo

List of contributors

John Dransfield
38A Tilehouse Green Lane, Knowle, West Midlands B93 9EY, UK

John Lay
RMC Aggregates (UK) Ltd, RMC House, Church Lane, Bromsgrove, Worcestershire
B61 8RA, UK

Robert Lewis
Elkem Materials Ltd, Elkem House, 4a Corporation Street, High Wycombe
HP13 6TQ, UK

Graeme Moir
East View, Crutches Lane, Higham, Near Rochester, Kent ME2 3UH, UK

Mark Murrin-Earp
RMC Aggregates (UK) Ltd, RMC House, Church Lane, Bromsgrove, Worcestershire
B61 8RA, UK

John Newman
Department of Civil Engineering, Imperial College, London SW7 2BU, UK

Phil Owens
Rosebank, Donkey Lane, Tring, Hertfordshire HP23 4DY, UK

Alan Poole
Parks House, 1D Norham Gardens, Oxford OX2 6PS, UK

Ray Ryle
3 Captains Gorse, Upper Basildon, Reading RG8 8SZ, UK

Karen Scrivener
EPFL, Swiss Federal Institute of Technology, Ecublens, Lausanne CH 1015, Switzerland

Lindon Sear
UK Quality ASL Association, Regent House, Bath Avenue, Wolverhampton, West Midlands WV1 4EG, UK

Ian Sims
STATS Consultancy, Porterswood House, Porterswood, St Albans AL3 6PQ, UK

Peter Wainwright
School of Civil Engineering, The University of Leeds, Leeds L52 9TT, UK

PART 1

Cements

1

Cements

Graeme Moir

1.1 Introduction

The aims and objectives of this chapter are to:

- describe the nature of Portland (calcium silicate-based) cements
- outline the manufacturing process and the quality control procedures employed
- review the cement hydration processes and the development of hydrated structures
- outline the influence of differences in cement chemistry and compound composition on the setting and strength development of concrete
- review cement types (including composite and masonry cements) and the nature of their constituents
- review the standards with which cements must comply and the applications for different cement types
- describe in outline methods used for chemical analysis and to study the hydration of cement
- briefly outline some health and safety aspects related to cement use

1.2 History of Portland cement manufacture

Portland cement is essentially a calcium silicate cement, which is produced by firing to partial fusion, at a temperature of approximately 1500°C, a well-homogenized and finely ground mixture of limestone or chalk (calcium carbonate) and an appropriate quantity of clay or shale. The composition is commonly fine tuned by the addition of sand and/or iron oxide.

The first calcium silicate cements were produced by the Greeks and Romans, who discovered that volcanic ash, if finely ground and mixed with lime and water, produced a hardened mortar, which was resistant to weathering. The reaction is known as the pozzolanic reaction and it is the basis of the contribution made to strength and concrete performance by materials such as fly ash, microsilica and metakaolin in modern concrete.

In the mid eighteenth century John Smeaton discovered that certain impure limes (these contained appropriate levels of silica and alumina) had hydraulic properties. That is, they contained reactive silicates and aluminates, which could react with water to yield durable hydrates, which resisted the action of water. Smeaton used this material in the mortar used to construct the Eddystone Lighthouse in 1759.

The term 'Portland cement' was first applied by Joseph Aspdin in his British Patent No. 5022 (1824), which describes a process for making artificial stone by mixing lime with clay in the form of a slurry and calcining (heating to drive off carbon dioxide and water) the dried lumps of material in a shaft kiln. The calcined material (clinker) was ground to produce cement. The term 'Portland' was used because of the similarity of the hardened product to that of Portland stone from Dorset and also because this stone had an excellent reputation for performance.

Joseph Aspdin was not the first to produce a calcium silicate cement but his patent gave him the priority for the use of the term 'Portland cement'. Other workers were active at the same time or earlier, most notably Louis Vicat in France. Blezard (1998) gives a comprehensive review of the history of the development of calcareous (lime-based) cements.

The cements produced in the first half of the nineteenth century did not have the same compound composition as modern Portland cements as the temperature achieved was not high enough for the main constituent mineral of modern cements, tricalcium silicate (C_3S), to be formed. The only silicate present was the less reactive dicalcium silicate (C_2S). The sequences of reactions, which take place during clinker production, are discussed in section 1.3.3.

The main technical innovations in cement manufacture which have taken place over approximately the last 150 years are summarized in Figure 1.1.

It was the introduction of the rotary kiln at the end of the nineteenth century that enabled a homogeneous product to be manufactured, which had experienced a consistently high enough temperature to ensure C_3S formation. During the twentieth century the nature of the product changed relatively little in terms of its overall chemistry and mineral composition but there have been considerable advances in production technology resulting in improved energy efficiency, improved quality control, reduced environmental impact and lower labour intensity.

It should be noted that the introduction of rotary kiln technology in the early twentieth century coincided with the publication of cement standards in the UK and the USA. Both standards required the strength of a briquette of cement paste to reach minimum values at 7 and 28 days.

The control of clinker composition has advanced from the volume proportions arrived at by trial and error in the late nineteenth century to precise control using rapid X-ray fluorescence techniques. The continuous improvements in manufacturing methods and quality control combined with market competitive pressures have resulted in a fourfold increase in the 28-day strength given by a typical European Portland cement at 28 days since the late nineteenth century (Blezard, 1998). In Europe, this strength escalation has effectively been controlled by the introduction of cement standards with upper as well as lower strength limits. The European Standard for Common Cements (EN 197-1) is outlined in section 1.6.3.

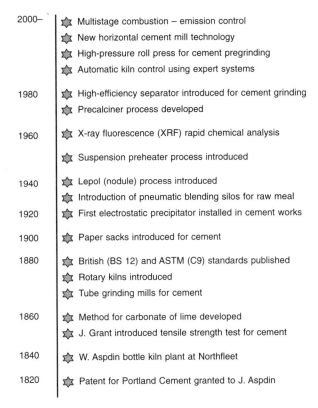

2000–	Multistage combustion – emission control
	New horizontal cement mill technology
	High-pressure roll press for cement pregrinding
	Automatic kiln control using expert systems
1980	High-efficiency separator introduced for cement grinding
	Precalciner process developed
1960	X-ray fluorescence (XRF) rapid chemical analysis
	Suspension preheater process introduced
1940	Lepol (nodule) process introduced
	Introduction of pneumatic blending silos for raw meal
1920	First electrostatic precipitator installed in cement works
1900	Paper sacks introduced for cement
1880	British (BS 12) and ASTM (C9) standards published
	Rotary kilns introduced
	Tube grinding mills for cement
1860	Method for carbonate of lime developed
	J. Grant introduced tensile strength test for cement
1840	W. Aspdin bottle kiln plant at Northfleet
1820	Patent for Portland Cement granted to J. Aspdin

Figure 1.1 Landmarks in Portland cement production.

1.3 Chemistry of clinker manufacture

1.3.1 Raw materials

Cement making is essentially a chemical process industry and has much in common with the manufacture of so-called heavy chemicals such as sodium hydroxide and calcium chloride. Close control of the chemistry of the product is essential if cement with consistent properties is to be produced. This control applies not only to the principal oxides which are present but also to impurities, which can have a marked influence on both the manufacturing process and cement properties.

As illustrated in Figure 1.2, a chemical analysis of Portland cement clinker shows it to consist mainly of four oxides: CaO (lime), SiO_2 (silica), Al_2O_3 (alumina) and Fe_2O_3 (iron oxide). In order to simplify the description of chemical composition, a form of shorthand is used by cement chemists in which the four oxides are referred to respectively as C, S, A and F.

Expressing the chemical analysis in the form of oxides, rather than the individual elements of silicon (Si), calcium (Ca) etc., has the advantage that the analysis total should

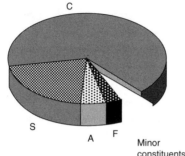

Main constituents	Cement chemist's shorthand	%
SiO$_2$	S	21.1
Al$_2$O$_3$	A	5.6
Fe$_2$O$_3$	F	3.0
CaO	C	65.5
		95.2

Minor constituents	%
Mn$_2$O$_3$	0.05
P$_2$O$_5$	0.15
TiO$_2$	0.30
MgO	1.50
SO$_3$(S)	1.20
Loss on ignition	0.50
K$_2$O	0.72
Na$_2$O	0.18
Fluorine	0.04
Chloride	0.02
Trace elements	0.01
	4.67

Figure 1.2 Typical chemical composition of Portland cement clinker.

come close to 100, and this provides a useful check for errors. Oxidizing conditions are maintained during the burning process and this ensures that the metallic elements present are effectively present as oxides although combined in the clinker as minerals.

The source of lime for cement making is usually limestone or chalk. As typically 80% of the raw mix consists of limestone, it is referred to as the primary raw material. The secondary raw material, which provides the necessary silica, alumina and iron oxide, is normally shale or clay. Small quantities of sand or iron oxide may be added to adjust the levels of silica and iron oxide in the mix. When proportioning the raw materials, an allowance must be made for ash incorporated into the clinker from the fuel that fires the kiln. Most cement plants worldwide use finely ground (pulverized) coal as the primary fuel. Increasingly, by-product fuels such as the residue from oil refining (petroleum coke) and vehicle tyres are being used to partially replace some of the coal.

Typical contents of the four principal oxides in a simplified cement making operation utilizing only two raw materials are given in Figure 1.3.

Note that the ratio of CaO to the other oxides is lower in the clinker than in the raw mix. This is a result of the incorporation of shale from the coal ash. The levels of the oxides are also increased as a result of decarbonation (removal of CO$_2$).

1.3.2 The modern rotary kiln

The rotary kilns used in the first half of the twentieth century were wet process kilns which were fed with raw mix in the form of a slurry. Moisture contents were typically 40% by mass and although the wet process enabled the raw mix to be homogenized easily, it carried a very heavy fuel penalty as the water present had to be driven off in the kiln.

In the second half of the twentieth century significant advances were made which have culminated in the development of the precalciner dry process kiln. In this type of kiln, the

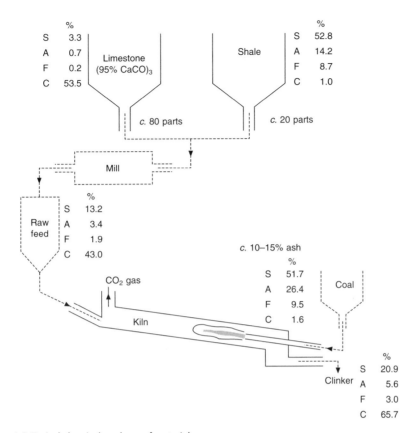

Figure 1.3 Typical chemical analyses of materials.

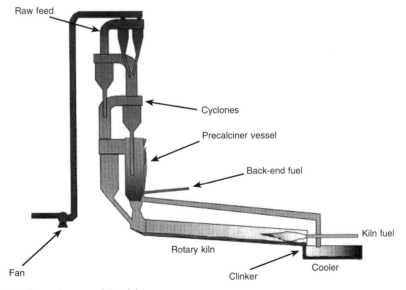

Figure 1.4 The modern precalciner kiln.

energy-consuming stage of decarbonating the limestone present in the raw mix is completed before the feed enters the rotary kiln. The precalcination of the feed brings many advantages, the most important of which is high kiln output from a relatively short and small-diameter rotary kiln. Almost all new kilns installed since 1980 have been of this type. Figure 1.4 illustrates the main features of a precalciner kiln.

The raw materials are ground to a fineness, which will enable satisfactory combination to be achieved under normal operating conditions. The required fineness depends on the nature of the raw materials but is typically in the range 10–30% retained on a 90 micron sieve. The homogenized raw meal is introduced into the top of the preheater tower and passes downwards through a series of cyclones to the precalciner vessel. The raw meal is suspended in the gas stream and heat exchange is rapid. In the precalciner vessel the meal is flash heated to ~900°C and although the material residence time in the vessel is only a few seconds, approximately 90% of the limestone in the meal is decarbonated before entering the rotary kiln. In the rotary kiln the feed is heated to ~1500°C and as a result of the tumbling action and the partial melting it is converted into the granular material known as clinker. Material residence time in the rotary kiln of a precalciner process is typically 30 minutes. The clinker exits the rotary kiln at ~1200°C and is cooled to ~60°C in the cooler before going to storage and then being ground with gypsum (calcium sulfate) to produce cement. The air which cools the clinker is used as preheated combustion air thus improving the thermal efficiency of the process. As will be discussed in section 1.5, the calcium sulfate is added to control the initial hydration reactions of the cement and prevent rapid, or flash, setting.

If coal is the sole fuel in use then a modern kiln will consume approximately 12 tonnes of coal for every 100 tonnes of clinker produced. Approximately 60% of the fuel input will be burned in the precalciner vessel. The high fuel loading in the static precalciner vessel reduces the size of rotary kiln required for a given output and also reduces the consumption of refractories. A wider range of fuel types (for example, tyre chips) can be burnt in the precalciner vessel than is possible in the rotary kiln.

Although kilns with daily clinker outputs of ~9000 tonnes are in production in Asia most modern precalciner kilns in operation in Europe have a production capability of between 3000 and 5000 tonnes per day.

1.3.3 Clinkering reactions and the minerals present in Portland cement clinker

Portland cement clinker contains four principal chemical compounds, which are normally referred to as the clinker minerals. The composition of the minerals and their normal range of levels in current UK and European Portland cement clinkers are summarized in Table 1.1.

It is the two calcium silicate minerals, C_3S and C_2S, which are largely responsible for the strength development and the long-term structural and durability properties of Portland cement. However, the reaction between CaO (lime from limestone) and SiO_2 (silica from sand) is very difficult to achieve, even at high firing temperatures. Chemical combination is greatly facilitated if small quantities of alumina and iron oxide are present (typically 5% Al_2O_3 and 3% Fe_2O_3), as these help to form a molten flux through which the lime and silica are able to partially dissolve, and then react to yield C_3S and C_2S. The sequence of reactions, which take place in the kiln, is illustrated in Figure 1.5.

Table 1.1 Ranges of principal minerals in European clinkers

Shorthand nomenclature	Chemical formula	Mineral name	Typical level (mass %)	Typical range (mass %)
C_3S	$3CaO \cdot SiO_2$ or Ca_3SiO_5	Alite	57	45–65
C_2S	$2CaO \cdot SiO_2$ or Ca_2SiO_4	Belite	16	10–30
C_3A	$3CaO \cdot Al_2O_3$ or $Ca_3Al_2O_6$	Aluminate	9	5–12
C_4AF	$4CaO \cdot Al_2O_3 \cdot Fe_2O_3$ or $Ca_4Al_2Fe_2O_{10}$	Ferrite	10	6–12

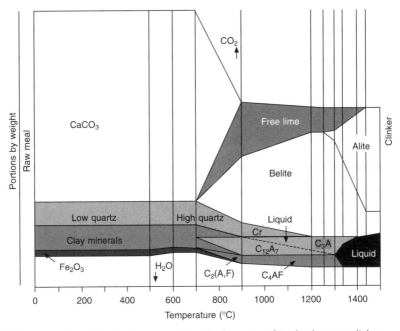

Figure 1.5 Sequence of reactions taking place during the formation of Portland cement clinker. (*Source:* Reproduced by courtesy of KHD Humbolt Wedag AG.)

The reaction requiring the greatest energy input is the decarbonation of $CaCO_3$, which takes place mainly in the temperature range 700–1000°C. For a typical mix containing 80% limestone the energy input to decarbonate the $CaCO_3$ is approximately 400 kCal/kg of clinker, which is approximately half of the total energy requirement of a modern dry process kiln.

When decarbonation is complete at about 1100°C, the feed temperature rises more rapidly. Lime reacts with silica to form belite (C_2S) but the level of unreacted lime remains high until a temperature of ~1250°C is reached. This is the lower limit of thermodynamic stability of alite (C_3S). At ~1300°C partial melting occurs, the liquid phase (or flux) being provided by the alumina and iron oxide present. The level of unreacted lime reduces as C_2S is converted to C_3S. The process will be operated to ensure that the level of unreacted lime (free lime) is below 3%.

Normally, C_3S formation is effectively complete at a material temperature of about 1450°C, and the level of uncombined lime reduces only slowly with further residence time. The ease with which the clinker can be combined is strongly influenced by the mineralogy of the raw materials and, in particular, the level of coarse silica (quartz) present. The higher the level of coarse silica in the raw materials, the finer the raw mix will have to be ground to ensure satisfactory combination at acceptable kiln temperatures.

Coarse silica is also associated with the occurrence of clusters of relatively large belite crystals around the sites of the silica particles. Figures 1.6(a) and 1.6(b) are photomicrographs of a 'normal' clinker containing well-distributed alite and belite and clinker produced from a raw meal containing relatively coarse silica.

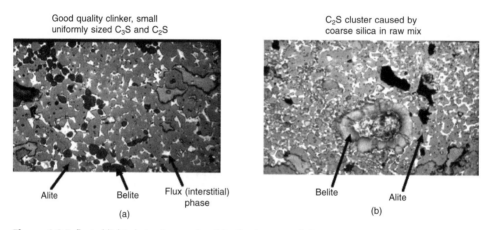

Figure 1.6 Reflected light photomicrographs of Portland cement clinker.

The belite present in the clusters is less reactive than small well-distributed belite and this has an adverse influence on cement strengths.

As the clinker passes under the flame it starts to cool and the molten C_3A and C_4AF, which constitute the flux phase, crystallize. This crystallization is normally complete by the time the clinker exits the rotary kiln and enters the cooler at a temperature of ~1200°C. Slow cooling should be avoided as this can result in an increase in the belite content at the expense of alite and also the formation of relatively large C_3A crystals which can result in unsatisfactory concrete rheology (water demand and stiffening).

1.3.4 The control ratios

The control of clinker composition, and optimization of plant performance is greatly assisted by the use of three ratios:

$$\text{Lime saturation factor LSF} = \frac{C}{2.8S + 1.2A + 0.65F} \times 100\%$$

$$\text{Silica ratio SR} = \frac{S}{A + F}$$

$$\text{Alumina ratio AR} = \frac{A}{F}$$

The most critical control ratio is the lime saturation factor, which is determined by the ratio of lime, to silica, alumina and iron oxide, and governs the relative proportions of C_3S and C_2S. The formula for LSF has been derived from high-temperature phase equilibria studies. When the LSF is above 100% there is an excess of lime, which cannot be combined no matter how long the clinker is fired, and this remains as free lime in the clinker. As a low level of uncombined lime must be achieved (~3% maximum and preferably below 2%), clinker LSFs normally lie in the range 95–98%. Figure 1.7 illustrates the influence of LSF on the content of C_3S and C_2S, and the firing temperature required to maintain clinker free lime below 2% (normally referred to as the combinability temperature). The contents of C_3S and C_2S have been calculated by the so-called Bogue method. This procedure is discussed further in section 1.3.5. Normally, if the LSF is increased at a particular cement plant, the raw mix must be ground finer, i.e. the percentage of particles coarser than 90 microns is reduced.

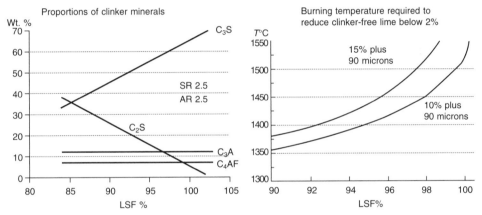

Figure 1.7 Influence of clinker LSF on compound composition and ease of combination.

In order to ensure optimum kiln performance and uniform cement quality it is essential that the LSF of the raw mix is maintained within a narrow band, ideally ±2% or, more precisely, with a standard deviation of better than 1%, determined on hourly samples. As a change in LSF of 1% (at constant free lime) corresponds to a change in C_3S of ~2% the C_3S variability range is approximately double that of the LSF range.

The silica ratio, SR, is the ratio of silica to alumina and iron oxide. For a given LSF, the higher the silica ratio, the more C_3S and the less C_3A and C_4AF will be produced. Of greater significance, with regard to clinker manufacture, is that the higher the SR, the less molten liquid, or flux, is formed. This makes clinker combination more difficult unless the LSF is reduced to compensate. The flux phase facilitates the coalescence of the clinker into nodules and also the formation of a protective coating on the refractory kiln lining. Both are more difficult to achieve as the SR increases.

The alumina ratio, AR, is normally the third ratio to be considered. An AR of ~1.4 is normally optimum for clinker burning, as at this value the quantity of liquid phase formed when partial melting first occurs at ~1300°C is maximized. High AR cements will have a high C_3A content, and this can be disadvantageous in certain cement applications, for example where it is desired to minimize the concrete temperature rise. Fortunately, the AR ratio is relatively easy to control by means of a small addition of iron oxide to the mix.

As a result of market requirements for cements from different sources to have similar properties and also to optimize clinker production there has been a trend to converge on a 'standard' clinker chemistry of

LSF   95–97%
SR   2.4–2.6
AR   1.5–1.8

At most plants the achievement of this ideal chemistry will require the use of corrective materials such as sand and iron oxide.

1.3.5 Calculation of clinker compound composition

The levels of the four clinker minerals can be estimated using a method of calculation first proposed by Bogue in 1929 (see Bogue, 1955). The method involves the following assumptions:

- all the Fe_2O_3 is combined as C_4AF
- the remaining Al_2O_3 (i.e. after deducting that combined in C_4AF) is combined as C_3A

The CaO combined in the calculated levels of C_3A and C_4AF and any free lime are deducted from the total CaO and the level of SiO_2 determines the proportions of C_3S and C_2S. The procedure can be expressed mathematically (in mass%) as follows:

$$C_3S = 4.071(\text{total CaO} - \text{free lime}) - 7.600SiO_2 - 6.718Al_2O_3 - 1.430Fe_2O_3$$

$$C_2S = 2.867SiO_2 - 0.7544C_3S$$

$$C_3A = 2.65Al_2O_3 - 1.692Fe_2O_3$$

$$C_4AF = 3.043Fe_2O_3$$

The calculated figures may not agree exactly with the proportions of the clinker minerals determined by quantitative X-ray diffraction or by microscopic point counting. However, they give a good guide to cement properties in terms of strength development, heat of hydration and sulfate resistance.

When calculating the compound composition of cements (rather than clinkers) the normal convention is to assume that all the SO_3 present is combined with Ca (i.e. is present as calcium sulfate). The total CaO is thus reduced by the free lime level and by $0.7 \times SO_3$. Examples of the calculation for cements are given in Table 1.7 later in this chapter.

1.3.6 Influence of minor constituents

As illustrated in Figure 1.2, approximately 95% of clinker consists of the oxides of CaO, SiO_2, Al_2O_3 and Fe_2O_3 (but present in combined form as the clinker minerals) and the remainder consists of the so-called minor constituents. The influence of minor constituents on cement manufacture and cement properties has been reviewed (Moir and Glasser, 1992; Bhatty, 1995).

Table 1.2 indicates the typical UK levels of the most commonly encountered minor constituents and summarizes their impact on the cement manufacturing process. The inputs of alkali metal oxides (Na_2O and K_2O), SO_3 and chloride have to be closely controlled because they are volatilized in the kiln and can cause severe operational problems associated with their condensation and the formation of build-ups in the kiln 'back end' and preheater.

Table 1.2 Influence of most commonly encountered minor constituents on manufacturing process

Minor constituent	Typical range of levels in UK clinkers	Influence on process
Na_2O	0.07–0.22	Alkali sulfates are volatilized in the kiln and condense in lower-temperature regions resulting in build-ups and blockages
K_2O	0.52–1.0	
SO_3	0.5–1.5	
Fluorine	0.01–0.20	Greatly assists combination by virtue of mineralizing action
Chloride	0.005–0.05	Alkali chlorides are highly volatile and cause build-ups and blockages
MgO	0.8–2.5	Slight fluxing action
Trace metals	5–100ppm	Slight – but some (e.g. thallium) have to be minimized to limit emissions to the environment

The alkali metals Na_2O and K_2O have a very strong affinity for SO_3 and a liquid phase containing Na^+, K^+, Ca^{2+} and SO_4^{2-} ions is formed which is immiscible with the main clinker liquid (molten C_3A and C_4AF). On cooling this crystallizes to yield alkali sulfates such as K_2SO_4, aphthitalite ($3K_2SO_4 \cdot Na_2SO_4$) and calcium langbeinite ($K_2SO_4 \cdot 2CaSO_4$). The crystallization products depend on the relative levels of the two alkali oxides and the level of SO_3. If there is insufficient SO_3 to combine with the alkali metal oxides then these may enter into solid solution in the aluminate and silicate phases. C_2S can be stabilized at temperatures above 1250°C thus impeding the formation of C_3S. A similar stabilization of C_2S, requiring 'hard burning' to lower the free lime level to an acceptable level, can occur if there is a large excess of SO_3 over alkalis.

A deficiency of SO_3 in the clinker is associated with enhanced C_3A activity and difficulties in achieving satisfactory early age concrete rheology.

Fluorine occurs naturally in some limestone deposits, for example in the Pennines in England, and has a beneficial effect on clinker combination. It acts as both a flux and mineralizer, increasing the quantity of liquid formed at a given temperature and stabilizing C_3S below 1250°C. The level in the clinker, however, must be controlled below ~0.25% in order to avoid a marked reduction in the early reactivity of cement.

Minor constituents also have to be controlled on account of their impact on cement properties and also concrete durability. Related to this, the levels of alkalis, SO_3, chloride and MgO are also limited by national cement standards or codes of practice. These aspects are reviewed in section 1.6.

1.4 Cement grinding

The clinker is normally conveyed to a covered store where, provided stocks are adequate, it will cool from the cooler discharge temperature of 50–80°C to a temperature approaching ambient. If clinker stocks are low then the clinker may be ground to cement without the opportunity to cool further during storage. The clinker is ground to a fine powder with approximately 5% calcium sulfate, which is added to control the early reactions of the aluminate phase. These reactions are considered in detail in section 1.5. The European standard for common cements also permits the addition of up to 5% of a minor additional constituent (mac), which, in practice, is normally limestone. Macs can be helpful in optimizing cement rheological properties.

The vast majority of cement produced throughout the world is ground in ball mills, which are rotating tubes containing a range of sizes of steel balls. A closed-circuit milling installation is illustrated in Figure 1.8.

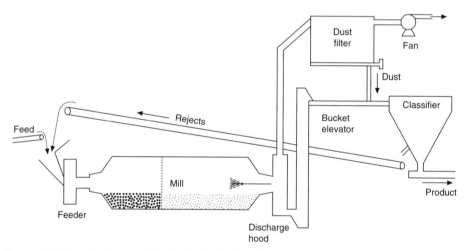

Figure 1.8 Schematic diagram of closed-circuit grinding mill.

Closed-circuit mills normally have two chambers separated by a slotted diaphragm, which allows the partially ground cement to pass through but retains the grinding balls. The first chamber contains large steel balls (60–90 mm in diameter), which crush the clinker. Ball sizes in the second chamber are normally in the range 19–38 mm.

The mill operates in a closed circuit in which the mill product passes to a separating device where coarse particles are rejected and returned to the mill for further grinding. The final product can thus be significantly finer than the material that exits the mill. Mills which do not have this separating stage are known as open-circuit mills and they are less efficient particularly at high cement finenesses (above 350 m^2/kg).

The efficiency of the clinker grinding process is very low. Less than 2% of the electrical energy consumed is used in actually fracturing the particles; the rest is converted to heat. Modern mills are equipped with internal water sprays, which cool the process by evaporation. Cement mill temperatures are typically in the range 110–130°C and at this temperature the hydrated form of calcium sulfate (gypsum, $CaSO_4 \cdot 2H_2O$) added to control the initial hydration reactions undergoes dehydration. This has some advantages but the level of dehydrated calcium sulfate has to be controlled to optimize the water demand properties of the cement. This aspect is discussed in section 1.5.5.

The low efficiency of the grinding process has resulted in considerable effort being directed to find more efficient processes. Some of these developments are listed in Figure 1.1. As a general rule the more efficient the grinding process, the steeper the particle size grading. The range of particle sizes is smaller and this can result in increased water demand of the cement, at least in pastes and rich concrete mixes. This is because with a narrow size distribution there are insufficient fine particles to fill the voids between the larger particles and these voids must be filled by water. Concerns over product performance in the market and also mechanical/maintenance problems with some of the new milling technologies have resulted in the ball mill retaining its dominant position. One compromise, which lowers grinding power requirement without prejudicing product quality, is the installation of a pre-grinder, such as a high-pressure roll press, to finely crush the clinker obviating the need for large grinding media in the first chamber of the ball mill.

1.5 Portland cement hydration

1.5.1 Introduction

The hydration of Portland cement involves the reaction of the anhydrous calcium silicate and aluminate phases with water to form hydrated phases. These solid hydrates occupy more space than the anhydrous particles and the result is a rigid interlocking mass whose porosity is a function of the ratio of water to cement (w/c) in the original mix. Provided the mix has sufficient plasticity to be fully compacted, the lower the w/c, the higher will be the compressive strength of the hydrated cement paste/mortar/concrete and the higher the resistance to penetration by potentially deleterious substances from the environment.

Cement hydration is complex and it is appropriate to consider the reactions of the silicate phases (C_3S and C_2S) and the aluminate phases (C_3A and C_4AF) separately. The hydration process has been comprehensively reviewed (Taylor, 1997).

1.5.2 Hydration of silicates

Both C_3S and C_2S react with water to produce an amorphous calcium silicate hydrate known as C–S–H gel which is the main 'glue' which binds the sand and aggregate particles together in concrete. The reactions are summarized in Table 1.3. C_3S is much more reactive than C_2S and under 'standard' temperature conditions of 20°C approximately half of the C_3S present in a typical cement will be hydrated by 3 days and 80% by 28 days. In contrast, the hydration of C_2S does not normally proceed to a significant extent until ~14 days.

The C–S–H produced by both C_3S and C_2S has a typical Ca to Si ratio of approximately 1.7. This is considerably lower than the 3:1 ratio in C_3S and the excess Ca is precipitated as calcium hydroxide (CH) crystals. C_2S hydration also results in some CH formation. The following equations approximately summarize the hydration reactions:

$$C_3S + 4.3H \Rightarrow C_{1.7}SH_3 + 1.3CH$$

$$C_2S + 3.3H \Rightarrow C_{1.7}SH_3 + 0.3CH$$

An important characteristic of C_3S hydration is that after an initial burst of reaction with water on first mixing it passes through a dormant, or induction, period where reaction appears to be suspended. This is of practical significance because it allows concrete to be placed and compacted before setting and hardening commences.

Table 1.3 Hydration of calcium silicates

Mineral	Reaction rate	Products of reaction
C_3S	Moderate	C–S–H with Ca:Si ratio ~ 1.7 CH (calcium hydroxide)
C_2S	Slow	C–S–H with Ca:Si ratio ~ 1.7 Small quantity of CH

Several theories have been developed to explain this dormant period. The most favoured is that the initial reaction forms a protective layer of C–S–H on the surface of the C_3S and the dormant period ends when this is destroyed or rendered more permeable by ageing or a change in structure. Reaction may also be inhibited by the time taken for nucleation of the C–S–H main product once water regains access to the C_3S crystals.

1.5.3 Hydration of C_3A and C_4AF

The reactions of laboratory-prepared C_3A and C_4AF with water, alone or in the presence of calcium sulfate and calcium hydroxide have been extensively studied (Odler, 1998). However, the findings should be interpreted with caution as the composition of the aluminate phases in industrial clinker differs considerably from that in synthetic preparations and hydration in cements is strongly influenced by the much larger quantity of silicates reacting and also by the presence of alkalis.

In the absence of soluble calcium sulfate C_3A reacts rapidly to form the phases C_2AH_8 and C_4AH_{19}, which subsequently convert to C_2AH_6. This is a rapid and highly exothermic reaction.

If finely ground gypsum ($CaSO_4 \cdot 2H_2O$) or hemihydrate ($CaSO_4 \cdot 0.5H_2O$) is blended with the C_3A prior to mixing with water then the initial reactions are controlled by the formation of a protective layer of ettringite on the surface of the C_3A crystals. The reaction can be summarized as:

$$C_3A + 3C + 3\bar{S} + 32H \Rightarrow C_3A \cdot 3C\bar{S} \cdot 32H$$

where in cement chemists' notation $\bar{S}$ represents SO_3 and H represents H_2O, i.e.

$$C_3A + \text{dissolved calcium } (Ca^{2+}) + \text{dissolved sulfate } (SO_4^{2-}) + \text{water} \Rightarrow \text{ettringite}$$

The more rapid dissolution of dehydrated forms of gypsum ensures an adequate supply of dissolved calcium and sulfate ions and will be more effective in controlling the reaction of finely divided or highly reactive forms of C_3A. The role of gypsum dehydration is considered further in section 1.5.5.

In most commercial Portland cements there will be insufficient sulfate available to sustain the formation of ettringite. When the available sulfate has been consumed the ettringite reacts with C_3A to form a phase with a lower SO_3 content known as monosulfate. The reaction can be summarized as:

$$C_3A \cdot 3C\bar{S} \cdot 32H + 2C_3A + 4H \Rightarrow 3(C_3A \cdot C\bar{S} \cdot 12H)$$

Many studies have shown that the hydration of C_4AF (or more correctly the $C_2A–C_2F$ solid solution) is analogous to that of C_3A but proceeds more slowly (Taylor, 1997). The iron enters into solid solution in the crystal structures of ettringite and monosulfate substituting for aluminium. In order to reflect the variable composition of ettringite and monosulfate formed by mixtures of C_3A and C_4AF they are referred to respectively as AFt (alumino-ferrite trisulfate hydrate) and AFm (alumino-ferrite monosulfate hydrate) phases. The hydration reactions of C_3A and C_4AF are summarized in Table 1.4.

Table 1.4 Hydration of aluminates

Mineral	Soluble calcium sulfate present	Reaction rate	Products of reaction
C_3A	No	Very rapid with release of heat	Hydrates of type C_2AH_8 and C_4AH_{19} which subsequently convert to C_2AH_6
C_3A	Yes	Initially rapid	Ettringite $C_3A3C\bar{S}\,32H$ which subsequently reacts to form monosulfate $3(C_3A \cdot C\bar{S} \cdot 12H)$
C_4AF	No	Variable (depends on Al/Fe ratio)	Hydrates of type $C_2(A,F)H_8$ and $C_4(A,F)H_x$ which subsequently convert to $C_3(A,F)H_6$ (hydrogarnet)
C_4AF	Yes	Variable but generally slow	Iron substituted ettringite (AFt) and subsequently iron substituted monosulfate (AFm)

1.5.4 Hydration of Portland cement

The hydration of Portland cement is rather more complex than that of the individual constituent minerals described above. A simplified illustration of the development of hydrate structure in cement paste is given in Figure 1.9.

When cement is first mixed with water some of the added calcium sulfate (particularly if dehydrated forms are present, and most of the alkali sulfates present (see section 1.3.5), dissolve rapidly. If calcium langbeinite is present then it will provide both calcium and sulfate ions in solution, which are available for ettringite formation.

The supply of soluble calcium sulfate controls the C_3A hydration, thus preventing a flash set. Ground clinker mixed with water without added calcium sulfate sets rapidly with heat evolution as a result of the uncontrolled hydration of C_3A. The cement then enters a dormant period when the rate of loss of workability is relatively slow. It will be more rapid, however, at high ambient temperatures (above 25°C).

Setting time is a function of clinker mineralogy (particularly free lime level), clinker chemistry and fineness. The finer the cement and the higher the free lime level, the shorter the setting time in general. Cement paste setting time is arbitrarily defined as the time when a pat of cement paste offers a certain resistance to penetration by a probe of

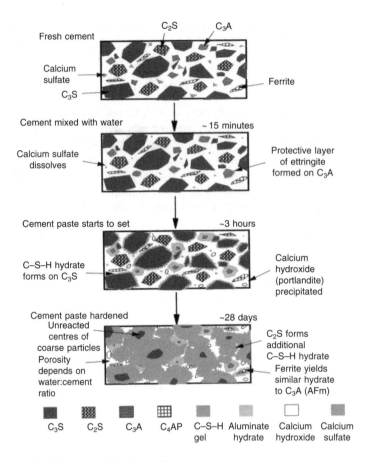

Figure 1.9 Simplified illustration of hydration of cement paste.

standard cross-section and weight (see section 1.7). Setting is largely due to the hydration of C_3S and it represents the development of hydrate structure, which eventually results in compressive strength.

The C–S–H gel which forms around the larger C_3S and C_2S grains is formed *in situ* and has a rather dense and featureless appearance when viewed using an electron microscope. This material is formed initially as reaction rims on the unhydrated material but as hydration progresses the anhydrous material is progressively replaced and only the largest particles (larger than ~30 microns) will retain an unreacted core after several years' hydration. This dense hydrate is referred to as the 'inner product'.

The 'outer hydration product' is formed in what was originally water-filled space and also space occupied by the smaller cement grains and by interstitial material (C_3A and C_4AF). When viewed using an electron microscope this material can be seen to contain crystals of $Ca(OH)_2$, AFm/AFt and also C–S–H with a foil- or sheet-like morphology. The structure of the outer product is strongly influenced by the initial water-to-cement ratio, which in turn determines paste porosity and consequently strength development.

The hydration of Portland cement involves exothermic reactions, i.e. they release heat. The progress of the reactions can be monitored using the technique of isothermal conduction calorimetry (Killoh, 1988).

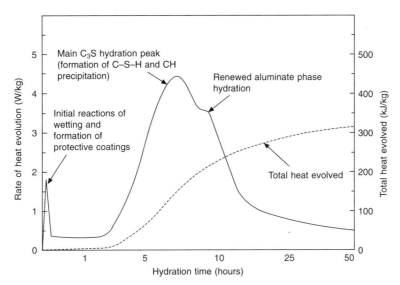

Figure 1.10 Heat of hydration of a cement paste determined by conduction calorimetry at 20°C.

The shoulder on the main hydration peak which is often seen at ~16 hours is associated with renewed ettringite formation which is believed to occur as a result of instability of the ettringite protective layer. In some cements with a low ratio of SO_3 to C_3A it may be associated with the formation of monosulfate.

The heat release is advantageous in cold weather and in precast operations where the temperature rise accelerates strength development and speeds up the production process. However, in large concrete pours the temperature rise, and in particular the temperature difference between the concrete core and the surface can generate stresses which result in 'thermal cracking'. Figure 1.11 illustrates the influence of concrete pour size on concrete temperature for a typical UK Portland cement. The data were obtained using the equipment described by Coole (1988).

The temperature rise experienced depends on a number of factors, which include:

- concrete placing temperature
- cement content
- minimum pour dimensions
- type of formwork
- cement type (fineness, C_3S and C_3A contents)

Cement heat of hydration (during the first ~ 48 hours) is highest for finely ground cements with a high C_3S content (>60%) and a high C_3A content (>10%).

By 28 days a typical Portland cement cured at 20°C can be expected to be ~90% hydrated. The extent of hydration is strongly influenced by cement fineness and in particular the proportions of coarse particles in the cement. Cement grains which are coarser than ~30 microns will probably never fully hydrate. Thus, cement particle size distribution has

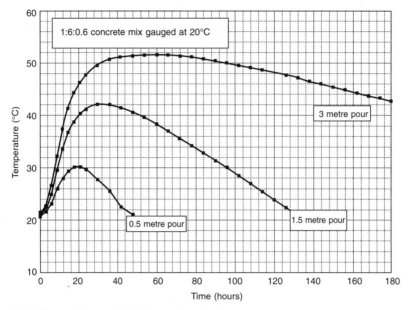

Figure 1.11 Influence of pour size (minimum dimensions) on concrete temperature.

a strong influence on long-term compressive strength. Cement produced in an open-circuit mill with a 45 micron sieve residue of 20% may give a 28-day strength ~10% lower than that of a cement produced from the same clinker but ground in a closed-circuit mill with a 45 micron sieve residue of 3% (Moir, 1994).

Elevated temperature curing, arising from either the semi-adiabatic conditions existing in large pours or from externally applied heat, is associated with reduced ultimate strength. This is believed to be due to a combination of microcracks induced by thermal stresses but also a less dense and 'well-formed' microstructure.

1.5.5 Optimization of level of rapidly soluble calcium sulfate

As described in section 1.3 cement mill temperatures normally lie in the range 100–130°C. Under these conditions the calcium sulfate dihydrate (gypsum) added to the mill undergoes dehydration first to hemihydrate ($CaSO_4 \cdot 0.5H_2O$) and then to soluble anhydrite ($CaSO_4$). These dehydrated forms of gypsum are present in commercial plasters and it is the formation of an interlocking mass of gypsum crystals which is responsible for the hardening of plaster once mixed with water.

The dehydrated forms of gypsum dissolve more rapidly than gypsum and this is beneficial in ensuring that sufficient Ca^{2+} and SO_4^{2-} ions are available in solution to control the initial reactivity of C_3A by forming a protective layer of ettringite. An inadequate supply of soluble calcium sulfate can result in a rapid loss of workability known as flash set. This is accompanied by the release of heat and is irreversible.

However, if too high a level of dehydrated gypsum is present, then crystals of gypsum crystallize from solution and cause a plaster or false set. This is known as false set because if mixing continues, or is resumed, the initial level of workability is restored.

This is because the gypsum needles which have developed a structure in the paste are broken and dispersed by re-mixing.

The cement manufacturer thus needs to optimize the level of dehydrated gypsum in the cement and match this to the reactivity of the C_3A present. This concept is illustrated in Figure 1.12.

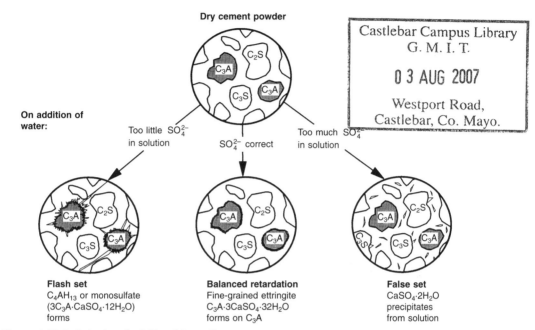

Figure 1.12 Optimization of soluble calcium sulfate.

Many natural gypsums contain a proportion of the mineral natural anhydrite ($CaSO_4$ – not to be confused with soluble anhydrite which is produced by gypsum dehydration). This form of calcium sulfate is unaffected by milling temperature and dissolves slowly in the pore solution providing SO_4^{2-} ions necessary for strength optimization but having no potential to produce false set.

Optimization of the level of readily soluble sulfate is achieved by a combination of:

- control of cement total SO_3 level
- controlling the level of natural anhydrite in the calcium sulfate used (either by purchasing agreement or by blending materials)
- controlling cement milling temperature

It must be emphasized that the optimum level of dehydrated gypsum can be influenced strongly by the use of water-reducing admixtures. Cement–admixture interactions are complex and some admixtures will perform well with certain cements but may perform relatively poorly with others.

1.5.6 Techniques used to study hydration

The following techniques are used to study the hydration of cement:

- thermal analysis
- X-ray diffraction
- scanning electron microscopy

In order to study hydrated cement it is necessary to remove free water and arrest hydration. This is normally done by immersing the sample in acetone and drying it at low temperature (<50°C) in a partial vacuum.

Thermal analysis

The most commonly applied technique is thermo-gravimetric analysis. A small sample of hydrated cement is placed in a thermobalance and the change in weight recorded as the sample is heated at a controlled rate from ambient to ~1000°C. The technique enables the proportion of certain hydrates present, such as ettringite and $Ca(OH)_2$ to be determined quantitatively.

X-ray diffraction

This technique is rapid but provides limited information as many of the hydrates present, notably C–S–H gel and calcium aluminate monosulfate are poorly crystalline and give ill-defined diffraction patterns.

Scanning electron microscopy (SEM)

This is a powerful technique, particularly when the microscope is equipped with a microprobe analyser. It involves techniques akin to X-ray fluorescence to determine the chemical composition of hydrates in the field of view. The high resolution of the SEM enables the microstructure of the hydrated cement paste in concrete or mortar to be studied. However, caution must be exercised when interpreting the images as specimen preparation and the vacuum required by most microscopes can generate features, which are not present in the moist paste.

1.5.7 Constitution of hydrated cement paste

Using the techniques described above it is possible to determine quantitatively the phases formed as Portland cement hydrates (Taylor, Mohan and Moir, 1985). A typical example is given in Table 1.5. It can be seen that the dominant phase present (by volume) is C–S–H with approximately equal quantities of calcium hydroxide and monosulfate. These hydrate proportions are changed significantly when Portland cement is blended or interground with pozzolanas (such as fly ash) or granulated blastfurnace slag. These reactions and the hydration products are discussed briefly in section 1.6.2.

1.6 Portland cement types

1.6.1 Standards

All developed countries have their own national standards for cements. These standards define the permitted cement composition and set performance requirements for properties

Table 1.5 Constitution of hydrated Portland cement paste

Anhydrous cement				
	Weight %			
C_3S	63			
C_2S	13			
C_3A	9.5			
C_4AF	7.5			
$CaSO_4$	5			

Hydrated cement paste				
		Volume %		
		3 days	28 days	365 days
Unreacted	C_3S	18	8	5
	C_2S	7	5	2
	C_3A	4	0	0
	C_4AF	5	5	2
Hydrates	Calcium hydroxide [$Ca(OH)_2$]	13	18	18
	C–S–H gel	39	47	54
	Monosulfate ($C_3A \cdot CaSo_4 \cdot 12H_2O$	4	16	20
	Ettringite ($C_3A \cdot 3CaSO_4 \cdot 32H_2O$)	11	0	0

such as setting time and development of compressive strength. They also describe the test procedures to be used to determine cement composition and cement properties. Although it is now rather out of date, particularly in relation to the standards in place in European countries, the publication by Cembureau, (1991) provides a useful review of cement types produced around the world.

In the past, national cement standards in many countries have been strongly influenced by those developed in the UK and published by the British Standards Institution (BSI) and by those developed in the USA and published by the American Society for Testing and Materials (ASTM). In 1991 The BSI published revised cement standards, which were closely aligned with the draft European Standard for Common Cements. This European Standard (EN 197-1) was adopted in 2000 by the following European countries: Austria, Belgium, Denmark, Finland, France, Germany, Greece, Iceland, Ireland, Italy, Luxembourg, Norway, Portugal, Spain, Sweden, Switzerland and the United Kingdom. The objective of this standard (in common with standards for other materials) is to remove barriers to trade. In order to meet this objective, existing national standards in the above countries were withdrawn in 2002. It can be expected that this European standard and the supporting standards for test methods (EN 196) and for assessment of conformity (EN 197-2) will have a strong influence on national (or regional) cement standards in the future.

1.6.2 Main cement types

Generic types

Table 1.6 summarizes the main generic types of Portland cement produced around the world. The descriptions given in the table are very general. National (and regional) standards have specific limits for the contents of constituents and there may be several

Table 1.6 Main types of cement produced around the world

Type	Designation	Constituents (excluding minor additional constituent permitted in some countries)	Applications
Normal Portland	'Pure'	Clinker and calcium sulfate	All types of construction except where exposed to sulfates
Sulfate-resisting Portland	'Pure'	Low C_3A clinker and calcium sulfate	Where concrete is exposed to soluble sulfates
White Portland	'Pure'	Special low iron content clinker and calcium sulfate	For architectural finishes
Portland fly ash cement	Composite	Clinker, fly ash and calcium sulfate	All types of construction. Some countries recognize improved sulfate resistance and protection against alkali silica reaction if fly ash level above ~25%. Low heat properties at high fly ash levels
Portland slag cement	Composite	Clinker, granulated blastfurnace slag and calcium sulfate	All types of construction. Most countries recognize improved sulfate resistance if slag level above ~60%. Protection against alkali silica reaction recognized in some countries. Low heat properties at high slag levels
Portland limestone cement	Composite	Clinker, limestone of specified purity and calcium sulfate	In Europe all types of construction
Pozzolanic cement	Composite	Clinker, natural pozzolana and calcium sulfate	All types of construction. Sulfate resistance properties not normally recognized
Masonry cement	Composite	Clinker, limestone (or lime) and air-entraining agent	Brickwork, blockwork and rendering

different 'sub-types' of cement with different levels of power station fly ash and blastfurnace slag and even mixtures of slag, fly ash and limestone. In this chapter the term composite is applied to all cements containing clinker replacement materials (other than a minor additional constituent).

Types of 'pure' Portland cement

Table 1.7 compares the typical chemical composition of normal grey, sulfate-resisting and white Portland cements. Many countries have national specifications for sulfate-resisting Portland cement (e.g. BS 4027, ASTM C150 Type V, French type CPA-ES). The European Committee responsible for cement standardization has been unable to reach agreement on the maximum C_3A level required to ensure sulfate resistance and sulfate-resistant Portland cements are not recognized as a separate cement type in EN 197-1. The maximum permitted C_3A level (calculated by the Bogue method) in BS 4027 is 3.5% and in ASTM C 150 (Type V) 5%.

Sulfate-resisting clinker is normally produced in relatively short production runs by a

Table 1.7 Typical composition of grey, white and sulfate-resisting Portland cement

	Grey %	Sulfate-resisting %	White %
SiO$_2$	20.4	20.3	24.6
Insoluble residue (IR)	0.6	0.4	0.07
Al$_2$O$_3$	5.1	3.6	1.9
Fe$_2$O$_3$	2.9	5.1	0.3
CaO	64.8	64.3	69.1
MgO	1.3	2.1	0.55
SO$_3$	2.7	2.2	2.1
Na$_2$O	0.11	0.10	0.14
K$_2$O	0.77	0.50	0.02
Loss ignition (LOI)	1.3	1.3	0.85
Free lime %	1.5	2.0	2.3
LSF %	96.6	97.3	95
SR	2.55	2.33	11.2
AR	1.76	0.71	6.3
Calculated mineral composition			
C$_3$S	57	62	67
C$_2$S	16	12	20
C$_3$A	9	0.9	4.5
C$_4$AF	9	16	0.9

rotary kiln, which also produces conventional grey clinker. The frequent changeovers disrupt the normal production process and have an adverse effect on the lifetime of the refractory bricks, which line the kiln.

Sulfate-resisting concrete can also be produced by ensuring an appropriate level of fly ash or blastfurnace slag. This can be achieved either by purchasing a factory-produced cement or (where national provisions permit) by blending fly ash or ground slag with cement. The current UK codes (BS 8500: 2002 Part 2) require a minimum of 25% fly ash or 70% blastfurnace slag, by weight of the combination.

The production of white cement clinker requires careful selection of materials and fuels to ensure the minimum content of iron oxide and of other colouring oxides such as chromium, manganese and copper. In order to achieve the best possible colour the clinker is normally fired under conditions where there is a slight deficiency of oxygen resulting in reduction of the colouring oxides to lower oxidation states, which have a lesser detrimental effect, and the clinker is quenched rapidly with water to prevent oxidation. All these measures increase the production cost and white cement sells at a significant premium over grey Portland cement.

Composite cements

Composite cements are cements in which a proportion of the Portland cement clinker is replaced by industrial by-products, such as granulated blastfurnace slag (gbs) and power station fly ash (also known as pulverized-fuel ash or pfa), certain types of volcanic material (natural pozzolanas) or limestone. The gbs, fly ash and natural pozzolanas react with the hydration products of the Portland cement, producing additional hydrates, which make a positive contribution to concrete strength development and durability. (Massazza, 1988; Moranville-Regourd, 1988).

In contrast, finely ground limestone, while not hydraulically active, modifies the hydration of the clinker minerals. It is introduced to assist in the control of cement strength development and workability characteristics. In some European countries, notably Belgium, France, the Netherlands and Spain, the quantity of composite cements produced considerably exceeds that of 'pure' Portland cement. According to statistics supplied by the association of European cement manufacturers (Cembureau) the average proportion of 'pure' Portland cement delivered in European Union countries was 38% in 1999 the remainder being 'composite cements'. The proportion of composite cements in the UK is at present very much lower, and the UK also differs from most European countries in that the addition of ground granulated blastfurnace slag (ggbs) and fly ash at the concrete mixer is well established. Further details are given in section 1.6.4.

Currently (2002), although there are regional variations, most of the ready-mixed concrete produced in the UK contains either ggbs (at ~50% level) or fly ash (at ~30% level).

The partial replacement of energy-intensive clinker by an industrial by-product or a naturally occurring material not only has environmental advantages but also has the potential to produce concrete with improved properties including long-term durability.

The characteristics of the constituents of composite cements are summarized in Table 1.8.

Table 1.8 Nature of composite cement constituents

Constituents	Siliceous fly ash (bituminous)	Natural pozzolana	Granulated blastfurnace slag	Limestone
Reaction type	Pozzolanic		Latently hydraulic	Hydration modifier
Brief description	Partly fused ash from the combustion of pulverized coal in power stations	Material of volcanic origin such as ash	Produced by rapid quenching of molten blastfurnace slag	Calcium carbonate of specified purity
Typical composition	Range	Range	Range	Typical
SiO_2	38–64	60–75	30–37	3
Al_2O_3	20–36	10–20	9–17	0.5
Fe_2O_3	4–18	1–10	0.2–2	0.5
CaO	1–10	1–5	34–45	51
MgO	0.5–2	0.2–2	4–13	2
S	–	–	0.5–2	–
SO_3	0.3–2.5	<1	0.05–0.2	0.3
LOI	2–7	2–12	0.02–1	42
K_2O	0.4–4	1–6	0.3–1	0.1
Na_2O	0.2–1.5	0.5–4	0.2–1	0.02
Reactive phases	Low lime silicate glass	Low lime silicate glass. Occasionally zeolite type material	High lime silicate glass	Calcium carbonate

Pozzolanic materials contain reactive (usually in glassy form) silica and alumina, which are able to react with the calcium hydroxide released by hydrating cement, to yield additional C–S–H hydrate and calcium aluminate hydrates. This reaction is much slower than the hydration of the clinker silicates and the strength development of pozzolanic cements is slower than that of 'pure' Portland cements. Provided moist curing is maintained, ultimate strengths may be higher than those of 'pure' Portland cement concrete. The C–S–H formed has a lower Ca:Si ratio than that found in pure Portland cement (~1.2 cf 1.7).

The nature of additions and the factors determining their performance are reviewed in detail in Chapter 3.

Fly ash has a significant advantage over natural pozzolanas as a result of the spherical shape of the glassy particles. These normally have a positive influence on concrete workability enabling concrete water contents to be reduced and thus offsetting the early age strength reduction. Fly ash performance may be improved by either removing coarse particles (using a classifier similar to that employed in closed-circuit cement grinding) or by co-grinding the ash with clinker. A British Standard (BS 3892: Part1) requires the ash to have a maximum 45 micron residue of 12% and ash meeting this standard is invariably classified. Blastfurnace slag is latently hydraulic and only requires activation by an alkaline environment to generate C–S–H and calcium aluminate hydrates. Although strength development is slower than with 'pure' Portland cement it is practicable to produce cements (or concretes) containing 50% slag (by weight of binder) with the same 28-day strength. Early strengths are considerably lower and those at 3 days may be approximately half those of 'pure' Portland cement (at 20°C).

The 'secondary' hydrates from the pozzolanic and latent hydraulic reaction develop a cement paste phase which has a lower permeability to water and to potentially aggressive ions such as sulfates and chlorides and this can have a positive impact on concrete durability.

While limestone constituents do not contribute significantly to strengths at 28 days, they do accelerate Portland cement hydration, and the reduction in early strength is normally less than in the case of fly ash. The most important characteristic of a limestone constituent is that it should comply with the purity requirements of the relevant standard (BS EN 197-1). Portland limestone cements with levels of limestone as high as 30% have been produced in France and Italy for many years and are used in a wide range of construction applications.

Figure 1.13 compares the strength development properties of cement prepared by blending a Portland cement with 20% ground limestone, granulated slag, natural pozzolana and fly ash (Moir and Kelham, 1997).

In Figure 1.13 the levels of slag and fly ash are lower than those found in UK cements or concretes but maintaining a constant addition level enables the reactivity of the materials to be compared directly.

1.6.3 The European Standard for Common Cements (EN 197-1)

As described above, CEN member countries voted to adopt EN 197-1 in 2000. In 2002 'conflicting' British Standards (such as BS 12) will be withdrawn. The British Standard for sulfate-resisting cement, BS 4027, will continue until such time as agreement is reached on a European Standard for sulfate-resisting cement. Table 1.9 summarizes the range of cement compositions permitted by EN 197-1.

While these are 'common cements' they are not all available in all CEN member countries. For example, Portland burnt shale cement requires a particular shale type, which is only found in southern Germany.

As well as defining the range of permitted cement compositions EN 197-1 also defines:

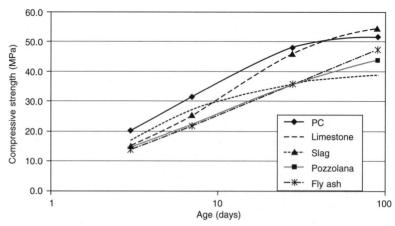

Figure 1.13 Comparative strength growth at 20°C of cements containing 20% limestone, slag, fly ash and pozzolana (BS 4550 concrete, w/c 0.60).

Table 1.9 Cement types and compositions permitted by EN 197-1

Cement type	Notation	Clinker %	Addition %
CEM I	Portland cement CEM I	95–100	–
CEM II	Portland-slag cement II/A-S	80–94	6–20
	II/B-S	65–79	21–35
	Portland-silica fume cement II/A-D	90–94	6–10
	Portland-pozzolana cement II/A-P	80–94	6–20
	II/B-P	65–79	21–35
	II/A-Q	80–94	6–20
	II/B-Q	65–79	21–35
	Portland-fly ash cement II/A-V	80–94	6–20
	II/B-V	65–79	21–35
	II/A-W	80–94	6–20
	II/B-W	65–79	21–35
	Portland-burnt shale cement II/A-T	80–94	6–20
	II/B-T	65–79	21–35
	Portland-limestone cement II/A-L	80–94	6–20
	II/B-L	65–79	21–35
	II/A-LL	80–94	6–20
	II/B-LL	65–79	21–35
	Portland-composite cement II/A-M	80–94	6–20
	II/B-M	65–79	21–35
CEM III	Blastfurnace cement III/A	35–64	36–65
	III/B	20–34	66–80
	III/C	5–19	81–95
CEM IV	Pozzolanic cement IV/A	65–89	11–35
	IV/B	45–64	36–55
CEM V	Composite cement V/A	40–64	36–60
	V/B	20–38	61–80

Notes:
All cements may contain up to 5% minor additional constituent (mac)
CEM V/A Composite cement contains 18–30% blastfurnace slag
CEM V/B Composite cement contains 31–50% blastfurnace slag
Proportions are expressed as % of the cement nucleus (excludes calcium sulfate)

- the permitted chemical composition of the individual constituents
- the permitted nature of the minor additional constituents (essentially either one of the permitted main constituents or a material derived from the clinker-making process)
- the permitted level of additives (e.g. grinding aids) – maximum of 1% in total including less than 0.5% organic
- minimum and maximum strengths for different strength classes
- minimum setting time
- chemical requirements (e.g. maximum SO_3, MgO, loss on ignition (LOI), insoluble residue (IR))
- conformity criteria to demonstrate compliance with EN 197-1.

Table 1.10 summarizes the compositional requirements for the constituents of EN 197-1 cements.

Table 1.10 Compositional requirements for constituents of EN 197-1 cements

Constituent	Requirements
Clinker	Minimum of 2/3rds C_3S plus C_2S. Ratio of CaO/SiO_2 >2. MgO < 5%
Blastfurnace slag	Minimum of 2/3rds $CaO + MgO + SiO_2$. Ratio $(CaO + MgO)/SiO_2$ > 1. Minimum 2/3rds glass
Siliceous fly ash	LOI < 7% (provided performance requirements for durability and admixture compatibility are met, otherwise <5%). Reactive CaO < 10% and free CaO < 1%. Free lime up to 2.5% permitted if soundness test passed
Calcareous fly ash	LOI < 7% (provided performance requirements for durability and admixture compatibility are met, otherwise <5%). Reactive CaO >10%. If reactive CaO >10% and <15% then reactive SiO_2 should be > 25%
Natural pozzolana	Reactive SiO_2 > 25%
Burnt shale	Compressive strength of >25 MPa at 28 days when tested according to EN 196-1. Expansion of <10 mm when blended with 70% cement
Limestone	$CaCO_3$ > 75%. Methylene blue adsorption (clay content) < 1.2g/100g. Total organic carbon < 0.2% (LL) or 0.5% (L)
Silica fume	LOI < 4%. Specific surface (BET) > 15.0 m^2/g

The compositional requirements for blastfurnace slag are the same as those in the British Standard for Portland blastfurnace cement (BS 146) and for ground granulated blastfurnace slag (BS 6699). Note that in EN 197-1 no test method is specified to determine the minimum glass content of 2/3rds.

Similarly the requirements for fly ash are essentially the same as those in the British Standards for Portland pulverized-fuel ash cement (BS 6588) and for pulverized-fuel ash (BS 3892 Part 1) although there are minor differences related to maximum LOI and CaO content.

Standards for concrete additions are reviewed in greater detail in Chapter 3.

Table 1.11 summarizes the mechanical and physical requirements for the strength classes permitted by EN 197-1. Compressive strength is determined using the EN 196-1 mortar prism procedure, which is outlined in section 1.7. Setting times are determined by the almost universally applied Vicat needle procedure and soundness by the method first developed by Le Chatelier in the nineteenth century. These methods are described in EN 196-3.

Table 1.11 Strength setting time and soundness requirements in EN 197-1

Strength class	Compressive strength MPa (EN 196-1 method)				Initial setting time (mm)	Soundness (expansion) (mm)
	Early strength		Standard strength			
	2 days	7 days	28 days			
32,5 N	–	≥16,0	≥32,5	≤52,5	≥75	
32,5 R	≥10,0	–				
42,5 N	≥10,0	–	≥42,5	≤62,5	≥60	≤10
42,5 R	≥20,0	–				
52,5 N	≥20,0	–	≥52,5		≥45	
52,5 R	≥30,0	–				

Currently, in the UK (2002) all bulk cement supplied is class 42,5 or 52,5 and the vast majority is pure Portland type (CEM I using BS EN 197-1 terminology). Less than 5% of the cement supplied is 'composite' and includes fly ash or limestone introduced during manufacture.

In contrast, in EU member countries (1999 data) the proportion of the three main strength classes was:

Class	Production (% of total tonnage)
32,5	48%
42,5	42%
52,5	10%

The proportion of the different cement types was as follows:

Type	Production (% of total tonnage)
CEM I Portland	38%
CEM II Portland composite	49%
CEM III Blastfurnace/slag	7%
CEM IV Pozzolanic	5%
CEM V Composite (and others)	1%

Thus, approximately 50% of the cement supplied was CEM II composite of which the highest proportion was Portland limestone cement (40% of the CEM II and 20% of the total tonnage).

In the UK the established practice is to add ground-granulated slag (to BS 6699) or pulverized-fuel ash (to BS 3892 Part 1) direct to the concrete mixer and to claim equivalence to factory-produced cement. The procedures required to demonstrate equivalence are described in BS 5328, which will be replaced, by BS EN 206-1 and the UK complementary standard, BS 8500 on 1 December 2003. In addition, some UK cement standards include cements with strength classes and properties outside the scope of BS EN 197-1 for common cements. For example, BS 146:2002, includes a low early strength class (L) for blastfurnace slag cements and there is a low 28-day class (22,5) in BS 6610:1996 for pozzolanic fly ash cement. Both of these standards will be withdrawn when European Standards covering the same scope are eventually published.

The chemical requirements of EN 197-1 cements are summarized in Table 1.12.

Table 1.12 Chemical requirements of EN 197-1 cements

Property	Cement type	Strength class	Requirements
Loss on ignition	CEM I (pure Portland) CEM III (>35% slag)	All	≤5.0%
Insoluble residue	CEM I		
	CEM III (.>35% slag)	All	≤5.0%
Sulfate (as SO_3)		32,5 N	
	CEM I	32,5 R	≤3.5%
	CEM II Portland composite cement	42,5 N	
	CEM IV Pozzolanic cement	42,5 R	
	CEM V Composite cement	52,5 N	≤4.0%
		52,5 R	
	CEM III (> 35% slag)	All	
Chloride	All	All	≤0.10%
Pozzolanicity	CEM IV Pozzolanic cement	All	Satisfies the test (EN 196–5)
MgO	All	All	≤5.0% in clinker

Upper limits for loss on ignition (LOI) and insoluble residue (IR) have featured in cement standards since their first introduction. The LOI ensured cement freshness and the IR limit prevented contamination by material other than calcium sulfate and clinker. However, the option to introduce up to 5% minor additional constituent (mac) has eroded their relevance. A higher level of assurance of consistency of performance is provided by the much more rigorous performance tests, which must be performed on random despatch samples at least twice per week.

The upper limit for SO_3 features in all cement standards and its purpose is to prevent expansion caused by the formation of ettringite from unreacted C_3A once the concrete has hardened. This expansive reaction, which occurs at normal curing temperatures a few days after mortar or concrete is mixed with water, should be distinguished from the phenomenon of delayed ettringite formation (DEF). All Portland cement can exhibit expansion as a result of DEF if they are subjected to high initial curing temperatures (above 80°C) and subsequent moist storage conditions. The cement factors which increase the risk of DEF have been identified by Kelham (1996). The cement SO_3 level has a positive influence on cement strength development, particularly at early ages, and over the past 20 years there has been a trend to raise the upper limit.

The purpose of the upper chloride limit is to reduce the risk of corrosion to embedded steel reinforcement. Although the upper limit is 0.1%, for prestressed concrete applications a lower limit may be agreed with the supplier and this will be declared on documentation. The concrete producer must, of course, consider all sources of chloride (water, aggregates, cement and admixtures) when meeting the upper limit for chloride in concrete.

The MgO limit of 5% on clinker ensures that unsoundness (expansion) will not occur as a result of the delayed hydration of free MgO. When MgO is present above ~2% in clinker it occurs as crystals of magnesium oxide (periclase) and these react relatively slowly to form $Mg(OH)_2$ (brucite) the formation of which is accompanied by expansion. Some countries such as the USA have a higher MgO limit (6%) but include an accelerated expansion test in which a sample of cement paste is heated in an autoclave and the expansion must remain below 0.08%.

EN 197-1 also describes the testing frequencies and the method of data analysis required to demonstrate compliance with the requirements of the standard. Note that the values given in Tables 1.11 and 1.12 are not absolute limits. A given percentage of the results obtained on random despatch samples may lie above or below these values. For example, for compressive strength 10% of results may lie above the upper limit for strength in a particular strength class but only 5% below the lower limit. For the physical and chemical requirement 10% of results may lie outside the limits. The spot samples taken at the point of cement despatch are known as autocontrol samples and the test results obtained as autocontrol test results.

Certificates confirming compliance with the requirements of the standards can be issued by EU Notified Certification Bodies (e.g. BSI Product Services) who follow the procedures detailed in EN 197-2. As EN 197-1 is a harmonized standard, the certification body can issue EC certificates of conformity which permit the manufacturer to affix the CE marking to despatch documents and packaging. The CE marking indicates a presumption of conformity to relevant EU health and safety legislation and permits the cement to be placed on the single European market. Failure to consistently meet the requirements of the standard may result in withdrawal of the EC certificate.

1.6.4 Other European cement standards

The CEN committee responsible for the development of standards for cements and building limes (CEN TC 51) has produced a number of standards for construction binders, which are close to finalization and adoption by member countries these are:

prEN 413-1: Masonry cement
prEN 645: Calcium aluminate cement (see Chapter 2)
EN 197-1: prA1: Amendment to EN 197-1 to include low heat common cements
prEN 197-X: Low early strength blast furnace cements
prEN 14216: Very low-heat special cements
EN 459-1: Building lime (published)
EN 13282: Hydraulic road binders (published)

Progress in developing a standard for sulfate-resisting cement is difficult as a result of national differences of view concerning the maximum level of C_3A in CEM I cements and the effectiveness of fly ash in preventing concrete deterioration. The solution may be a laboratory performance test but difficulties have been experienced in achieving a satisfactory level of reproducibility.

1.6.5 Other national standards

While the new European Standards can be expected to have a strong influence on the future development of national standards around the world, ASTM-based standards are likely to remain important in many countries. The main differences between current ASTM cement standards and EN 197-1 are shown in Table 1.13.

Table 1.13 Comparison between ASTM standards and EN 197-1

	ASTM C 150	EN 197-1
Compressive strength	Minimum values only	Three main strength classes with upper and lower limits
Initial set	Min 45 mins all types	Class 32,5 min 75 mins Class 42,5 min 60 mins Class 52,5 min 45 mins
Minor additional constituent	Not permitted	Permitted (<5%)
SO_3	Type I $C_3A \leq 8\%$ max 3.0% Type I $C_3A > 8\%$ max 3.5% Slag and pozzolanic cements max 4.0%	32,5 and 42,5 N max 3.5% 42,5 R and 52,5 max 4.0%
Chloride	No limit	Max 0.1%
MgO	Max 6.0% in cement	Max 5.0% in clinker

1.7 Cement production quality control

Different customers have different priorities concerning the specific properties of greatest importance to their business and Table 1.14 attempts to summarize these. The 'cement property' of prime importance is consistency of performance in relation to the aspects of greatest importance to the customer.

Table 1.14 Cement quality aspects of importance to customers

Primary aspect (of importance to most customers)	Secondary aspects (can be important to some customers)
Strength at 28 days Early strength Water demand Strength growth (and correlation between early and late strengths) Alkali content consistency of these aspects	Performance with slag and ash Response to changes in w/c ratio Performance with admixtures Premature stiffening Delayed stiffening Heat of hydration Colour Flowability Temperature Storage properties Consistency of these aspects

While the cement manufacturer can control certain properties, such as alkali content during the manufacturing process it is not possible to control strength directly because of the time delay before results are obtained. In practice control over cement properties is achieved by control of

- raw mix chemistry, fineness and homogeneity
- clinker chemistry and degree of heat treatment (burning)
- cement fineness, SO_3 level and forms of SO_3 present.

The chemical composition of raw materials for cement making, clinkers and cements can be determined by so-called wet chemical analysis methods. (BS EN 196-2, 1995).

This involves dissolving the materials in acid, and using a range of analytical techniques which are specific to the elements to be determined. The analysis requires between 2 and 6 hours to complete, depending on the facilities available, to determine the levels of the oxides S, A, F and C. Wet techniques also require highly skilled staff, if reliable results are to be obtained.

Fortunately, a rapid method of analysis known as X-ray fluorescence became available in the late 1960s and this technique is almost universally applied at cement works around the world. Sample preparation and the principles of the technique are outlined in Figure 1.14.

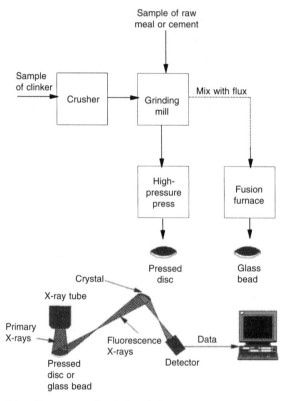

Figure 1.14 Outline of X-ray fluorescence chemical analysis.

The sample of raw meal, clinker or cement is inserted in the analyser, either in the form of a pressed disc of finely ground material or after fusing into a glass bead. In the analyser, the specimen is irradiated with X-rays, which cause secondary radiation to be emitted from the sample. Each chemical element present emits radiation of a specific frequency, and the intensity of the radiation is proportional to the quantity of that element present in the sample.

All UK cement plants (and all modern cement plants around the world) are equipped with an X-ray fluorescence analyser and use the technique to determine and control the levels of CaO, SiO_2, Al_2O_3, Fe_2O_3, MgO, K_2O, Na_2O, SO_3 and Cl in cement raw meal and clinker. The current cost of manually operated equipment is ~£150 000.

There is a trend towards continuous on-line analysis of raw materials either before or just after the raw mill using the technique of prompt gamma ray neutron activation analysis (PGNAA), (Macfadyen, 2000). There is also a trend towards fully automated laboratories where the processes outlined in Figure 1.13 are undertaken by a robot, similar to those used, for example, in car assembly.

The level of uncombined lime (free lime) present in clinkers is normally determined by extracting the CaO into hot ethylene glycol and titrating the solution with hydrochloric acid. Free lime can also be determined using the technique of X-ray diffraction and this is finding increasing favour as it is easier to automate than the glycol extraction method.

Close control over cement milling is essential to ensure a product with consistent properties. Although cement particle size distribution can be determined directly using laser diffraction techniques there have been difficulties with stability of the measurements over a period of time and the main methods most commonly used on cement plants for fineness determination are surface area (SA) determined by air permeability and sieving. In the UK the cement surface area is determined by placing a precise weight of cement in a 'Blaine' cell, which is then compacted to a constant volume. The time taken for a fixed volume of air to pass through the cell is a function of the cement surface area. The proportion of coarse particles present, as indicated by the 45-micron (or for finely ground cements 32-micron) sieve residue has a greater influence on cement 28-day strength than the surface area and it is important to monitor this parameter closely. The technique most commonly use is air-jet sieving where a stream of air passes through the sieve, agitating the material above the sieve and greatly speeding up the passage of sub-sized particles through the sieve. Instrumental techniques to measure cement fineness are improving and the trend is towards 'on-line' sampling and analysis linked to automatic mill control.

In European plants, cement strengths are determined using the EN 196-1 mortar test procedure. This utilizes a mortar, which consists of 3 parts (by weight) sand to 1 part cement. The dry sand, which is certified as meeting the requirements of the standard, is supplied in pre-weighed plastic bags, which simplifies the batching and mixing procedure. The w/c ratio is fixed at 0.50 for all cement types and the mortar is placed in a mould, which yields three prisms of dimensions $40 \times 40 \times 160$ mm. Compaction is either by vibration or by 'jolting' on specially designed tables. The prisms can be broken in flexure prior to compressive strength testing but this stage is normally omitted and the prisms simply snapped in two using a simple, manually operated, breaking device. The broken ends of each prism are tested for compressive strength using platens of dimensions 40×40 mm. Normally three prisms are broken at each test age and the result reported is the mean of six tests. A typical class 42,5 cement will give a strength of 55–59 MPa at 28 days. This is a similar strength level to that obtained if the cement is tested in a crushed granite concrete (such as the materials used in BS 4550 testing) with a w/c of 0.50. Most UK cement plants should achieve an annual average standard deviation for the 28-day strength results of main products in the range 1.5–2.0 MPa.

The setting time of cement paste is determined using the Vicat needle according to EN 196-3. This requires frequent testing of penetration in order not to 'miss' the initial and final set. Consequently there is a trend to introduce automation and equipment can be purchased which will determine the setting time of a number of cement samples (e.g. 12) simultaneously.

Effective quality control must start with the quarrying, proportioning and blending of the raw materials. If the raw mix is variable then no amount of adjustments to subsequent control parameters will eliminate product variability.

1.8 Influence of cement quality control parameters on properties

1.8.1 Key parameters

The cement properties of water demand (workability), setting behaviour and strength development are largely determined by the following 'key' cement quality control parameters:

- Cement fineness (SA and 45 micron residue)
- Loss on ignition (LOI)
- Clinker alkalis and SO_3
- Clinker-free lime
- Clinker compound composition (mainly calculated C_3S and C_3A levels)
- Cement SO_3 and the forms of SO_3 present

All of the above, with the exception of the forms of SO_3 present, can be determined rapidly by a competent cement plant laboratory. Specialist thermal analysis equipment is required to determine the forms of calcium sulfate present. Although some cement plants have this equipment it is more generally found at central laboratories.

1.8.2 Water demand (workability)

The initial hydration reactions and the importance of matching the supply of readily soluble calcium sulfate to clinker reactivity were reviewed in detail in section 1.5.5. In order to ensure satisfactory workability characteristics the cement producer should:

- maintain an appropriate ratio of SO_3 to alkalis in the clinker
- achieve a uniform and appropriate level of dehydrated gypsum by a combination of controlling the level of natural anhydrite in the 'calcium sulfate' used and the cement milling conditions

The partial replacement of clinker by fly ash should result in a reduction in water demand with a consequent increase in strength which partially offsets the lower early strengths associated with the lower reactivity of the ash. The reduced water demand can be mainly attributed to the glassy spheres present in fly ash which lubricate the mix. The lower density of the ash compared to cement also increases the cement paste volume. Blastfurnace slag grinds to yield angular particles. The slag is unreactive during initial hydration and generally has a neutral influence on water demand. Limestone can have a positive influence on water demand particularly when compared to a more coarsely ground pure Portland cement of the same strength class. This is because the fine limestone particles result in a more progressive (optimized) cement particle size distribution with a lower proportion of voids, which must be filled with water.

1.8.3 Setting time

Cement paste setting behaviour is strongly influenced by clinker reactivity and by cement fineness. The main factors, which tend to shorten setting time, are increases in the levels of:

- clinker-free lime
- cement fineness (surface area SA)
- C_3S content
- C_3A content

Because the setting time test is carried out on a sample of cement paste, which is gauged with water to give a standard consistency, any change in cement properties, which increases the water required, for example a steeper particle size distribution, will extend the setting time. The effects of a steeper size distribution are much more apparent in paste than in concrete and concretes made from the same cement may not exhibit the same magnitude of extension in setting time.

Setting time may also be extended by the presence of certain minor constituents. An example is fluorine. An increase in clinker fluorine level of 0.1% can be expected to increase setting time by ~ 60 minutes.

Both fly ash and slag will increase setting time while a Portland limestone cement may have a slightly shorter setting time than the corresponding pure Portland cement.

1.8.4 Strength development

Fineness

As discussed in section 1.7, on cement plants, cement fineness is normally determined with reference to surface area (SA) and 45-micron residue. While surface area is a good guide to the early rate of hydration of cement and thus early strengths, it is a less reliable guide to late strengths and, in particular, to 28-day strengths. This is because under standard curing conditions clinker particles which are coarser than approximately 30 microns are incompletely hydrated at 28 days. For a given SA the lower the 45-micron residue, the higher the 28-day strength. In EN 196-1 mortar an increase in 45-micron residue level of 1.0% can be expected to lower 28-day strength by ~ 0.4 MPa. In general, more modern milling installations, and in particular those with high efficiency separation, will yield cements with steeper size distributions. The steeper size distribution may require the introduction of a minor additional constituent (mac or filler) to control 28-day strength and thus remain within a certain strength class (Moir, 1994).

Loss on ignition (LOI)

Cements which contain up to 5% limestone minor additional constituent will have relatively high LOI values as a result of the decomposition of $CaCO_3$ during the test. In the case of cements which do not contain a limestone mac an increase in LOI above the 'base level' originating from residual water combined with calcium sulfate represents water and CO_2, which have combined with the clinker during storage or cement milling. Note that in some cases the base level may be slightly increased by $CaCO_3$ present as an impurity in the calcium sulfate.

The pre-hydration of clinker represented by the increased LOI has a marked influence on strength development. Typically, for a cement without a calcareous mac an increase in LOI of 1% can be expected to reduce EN 196-1 mortar 28-day strength by ~4 MPa.

The influence of LOI when a cement contains a calcareous mac is much less clear. The reduction in strength accompanying an increase in LOI of 1% associated with a higher filler level (1% LOI as CO_2 represents an addition of 2.5% high-grade $CaCO_3$) will result in a 28-day strength reduction of ~1 MPa.

In cement with several sources of LOI more sophisticated techniques (such as X-ray diffraction, or direct determination of carbon) can be used to determine the $CaCO_3$ level.

Clinker alkalis and SO₃

As described in section 1.3.6, the alkali metal oxides (Na_2O and K_2O) have a strong affinity for SO_3 and if sufficient SO_3 is present they will crystallize as alkali sulfates. In this form they are readily soluble and quickly dissolve in the gauging water modifying strength development properties. Soluble alkalis accelerate early strength development and depress late (28-day) strengths (Oesbaek, Jons, 1980).

Almost all CEM I type cements available in Europe will have eqNa$_2$O (calculated as $Na_2O + 0.66K_2O$) in the range 0.3–1.0%. Figure 1.15 illustrates the influence of cement alkalis (expressed as eqNa$_2$O) on the strength development of EN 196-1 mortar prisms. Note that in these clinkers there was sufficient SO_3 to combine most of the alkali present as sulfates.

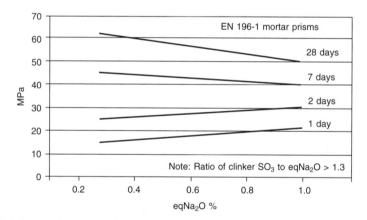

Figure 1.15 Influence of cement alkalis on strength development when present as soluble alkali sulfates.

A similar reduction in late strength can be achieved by adding alkali sulfates to cement; the increase in early strengths is normally less than when the sulfates are present in clinker. At early ages the dissolved alkali sulfate accelerates C_3S hydration. After approximately 1 day of hydration the supply of sulfate from calcium sulfate is exhausted and the pore solution consists essentially of Na^+, K^+ and OH^- ions whose pH is a function of the dissolved alkali. With high alkali cements the pH level may exceed 13.5 and this appears to be associated with inhibited silicate hydration.

In EN 196-1 mortar an increase in cement eqNa$_2$O (in soluble form) can be expected to increase 2-day strength by ~0.8 MPa and reduce 28-day strength by ~1.5 MPa.

Data for experimental clinker prepared with a very high level of clinker alkali and SO_3

show that the depression of 28-day strength remains almost linear but the acceleration of early strength reaches a maximum and a negative influence may be detected at very high levels. Figure 1.16 illustrates the results obtained. Note that eqNa$_2$O levels above 1% are unlikely to be encountered in Europe.

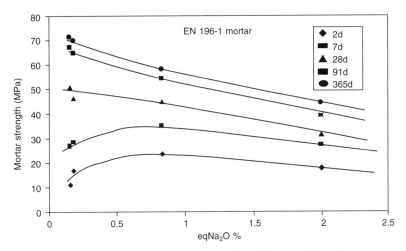

Figure 1.16 Influence of very high level of eqNa$_2$O combined in clinker as alkali sulfate.

If the level of clinker SO$_3$ is insufficient to combine with the alkali then the accelerating effect is reduced but the depression of 28-day strength is almost the same. This is because the alkali held in solid solution in the clinker minerals is released as hydration progresses.

Cements produced from high alkali clinker tend to give a better performance with slag and fly ash than equivalent low-alkali clinker. This is partly because the pH of the pore solution is reduced both by dilution and by absorption of alkalis in the lower calcium C–S–H formed (Taylor, 1987) but also because of the more aggressive attack on the glassy phases present in the slag and fly ash.

Free lime

Cement-free lime levels should normally lie in the range 0.5–3%. Late strengths are normally maximized by a low free lime level (as this maximizes the combined silicate content) but, as discussed in section 1.8.2, low free lime levels are associated with low reactivity and extended setting times. For cements with 'normal' parameters of LSF ~95 and SR 2.5 an increase of free lime level of 1% can be expected to reduce EN 196-1 mortar 28-day strength by ~1.5 MPa.

Compound composition

Strengths at ages of 1, 2 and 7 days are almost linearly related to C$_3$S content. At 28 days C$_2$S makes a significant contribution but the contribution does depend on C$_2$S reactivity, which is determined by clinker microstructure and by impurities present in the crystal lattice (solid solution effects). For example, the presence of belite clusters associated with coarse silica in the raw mix will reduce the contribution from C$_2$S at 28 days.

When clinkers were prepared in a batch rotary kiln with identical chemistries other than the ratio of C$_3$S to C$_2$S (Kelham and Moir, 1992) it was shown that with well-combined clinkers the highest 28-day strength was obtained with a moderate C_3S level. Results are illustrated in Figure 1.17.

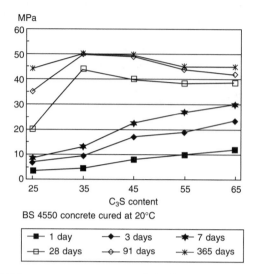

Figure 1.17 Influence of C_3S content on concrete strength development.

At 2 days and 7 days an increase in cement C_3S content of 10% can be expected to increase EN 196-1 mortar 2-day and 7-day strengths by ~3.5 and ~5 MPa respectively. The influence on 28-day strength is less certain but will normally be slightly positive.

For C_3A contents in the range 5–12% an increase in C_3A level will normally increase 28-day strength in standard quality control tests. This effect is more likely to be seen if the C_3A content increases as a result of an increase in AR rather than a decrease in SR. A decrease in SR will also reduce the total silicate content. The positive influence of C_3A is attributed to its higher reactivity compared to C_4AF, which results in a greater volume of hydrates and thus lower cement paste porosity at 28 days.

In field concretes (as opposed to fixed w/c laboratory mortars) the benefits of a higher C_3A content may be offset by a greater rate of slump loss resulting in a higher w/c ratio in order to maintain the required slump level.

SO_3 level and forms of SO_3 present

While the forms of calcium sulfate present can have a marked influence on the water demand of cement in concrete it is the total cement SO_3, which has the primary influence on strength properties. Anhydrite, if present, will dissolve during the first 24 hours of hydration and be available to participate in the strength-forming hydration reactions in the same manner as calcium sulfate from gypsum.

The response of a cement to a change in cement SO_3 level is influenced by a number of factors, which include:

- the alkali content and in particular the alkali sulfate (soluble alkali) content
- the C_3A level
- the cement fineness.

When optimizing the SO_3 level it is important to monitor the influence of the change on concrete water demand/rheology. It is also important to ensure that cements with different SO_3 levels are compared on a meaningful basis. For example, at constant SA the higher SO_3 cements will normally have higher 45-micron residues and consequently lower 28-day strengths.

Most clinkers will show a significant increase in early strength when the cement SO_3 level is increased from 2.5% to 3.5%. The influence on 28-day strength is generally much less but still positive. Typical results from laboratory tests are shown in Figure 1.18. In the example shown the adverse influence of cement SO_3 level on concrete slump could have been reduced by replacing a proportion of the gypsum by natural anhydrite.

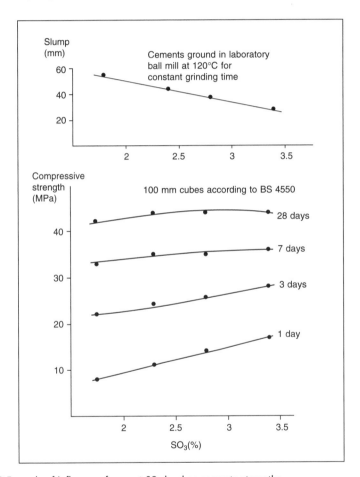

Figure 1.18 Example of influence of cement SO_3 level on concrete strengths.

In most countries the opportunity to optimize cement SO_3 is restricted by the upper limits for SO_3 in the relevant standards.

One important difference between a factory-produced composite cement and a concrete mixer blend is that in the latter it is not possible to control SO_3 level. With high slag blends in particular, the cement SO_3 level will be much lower than that of a factory-produced cement.

1.9 Relationship between laboratory mortar results and field concrete

Although UK cement producers replaced the BS 4550 concrete test with the EN 196-1 mortar test in 1991 as the primary assessment of cement strength (as required by BS 12: 1991) parallel testing in concrete has continued. A large quantity of data have been generated concerning the relationship between mortar and concrete strengths (Moir, 1999). Figure 1.19 shows the relationship between strength results obtained by the two methods.

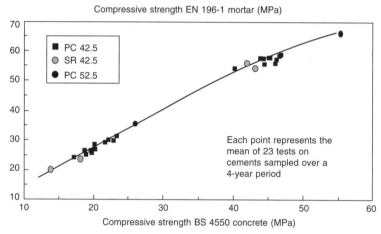

Figure 1.19 Relationship between EN 196-1 mortar and BS 4550 concrete strengths.

Although there is a clear 'average' relationship it is also clear that different cements have slightly different mortar to concrete (m/c) ratios and that this is a function of cement fineness and chemistry. This is not surprising as different cements respond differently to changes in cement content and w/c. This phenomenon is illustrated in Figure 1.20.

It can be seen that the 28-day strength ranking of the cements was different at the three different cement contents (and consequent w/c ratios). Thus no single test procedure, whether mortar or concrete, can reliably predict cement performance across a range of cement contents. The most important influence is cement alkali content as low alkali cement will tend to give relatively higher 28-day strengths at low w/c ratios and high alkali cements will tend to give lower strengths. Size distribution is also important as at high w/c ratios cement hydration will be more complete and the influence of coarse particles becomes more important.

1.10 Applications for different cement types

A detailed review of cement applications and the specification of concrete are outside the scope of this chapter. However, Table 1.15 summarizes the main applications likely to be selected for the principal generic cement types (and mixer blend equivalents) available in the UK. Note that this list has been prepared in 2002 and anticipates the publication of BS

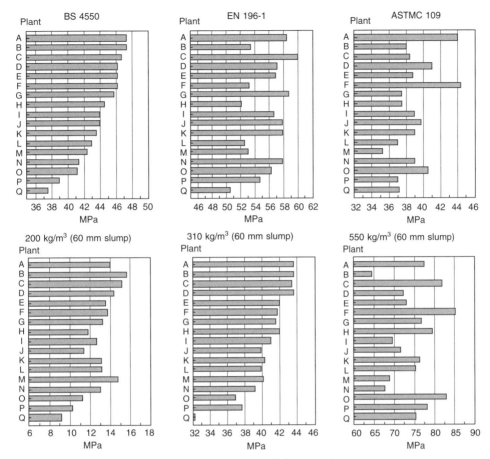

Figure 1.20 Influence of test procedure and w/c ratio on 28-day strengths.

EN 413-1. Additional cement types (drawn from the options available in BS EN 197-1) may become commercially significant in the future.

The choice of cement for structural concrete should be in accordance with BS 8500: 2002 after implementation on 1 December 2003 when it will supersede BS 5328. BS 8500 recognizes the sulfate-resisting properties of slag and fly ash concretes and the reduced risk of asr if appropriate levels of these materials are present.

It must be recognized that in many applications that cement choice has a much lesser influence on concrete long-term performance than the practical aspects of mix control, cement content, water content, aggregate quality, compaction, finishing and curing.

1.11 Health and safety aspects of cement use

Dry cement normally has no harmful effect on dry skin. However, when cement is mixed with water a highly alkaline solution is produced. The pH quickly exceeds 13 and is highly corrosive to skin.

Precautions should be taken to avoid dry cement entering the eyes, mouth and nose

Table 1.15 Applications for Portland cement types available in the UK

Cement type	BS	BS EN 196-1 designation	Clinker content*	Main applications
Portland cement	BS EN 197-1	CEM I 42,5 & 52,5 N	95–100	All types of general construction where resistance to sulfates or control of concrete temperature rise is not required.
Portland cement	BS EN 197-1	CEM I 52,5 R	95–100	As for 42.5 but where higher early strengths are required (for example, in precast concrete production or winter (concreting)
Packed Portland cement with integral air-entraining agent	BS EN 197-1	CEM I 32,5	95–100	Where improved concrete placing and finishing properties and resistance to freeze–thaw cycling are required without the need to use an air-entraining agent
Sulfate-resisting cement	BS 4027	Class 42.5	100	Where resistance to sulfates present in ground water is required. Alkali content <0.60% (optional) and can be used to reduce risk of alkali silica reaction
Masonry cement	BS EN 413-1	MC 22,5	≥40	For brick laying mortar and rendering. Gives improved plasticity and adhesion as well as improved water retention and durability
Portland fly ash cement	BS EN 197-1	CEM II/B-V	65–79	All types of general construction. When fly ash level exceeds 25% gives sulfate resistance and protection against alkali silica reaction. Moderate low heat at upper end of fly ash range (max 35%)
Pozzolanic cement	BS EN 197-1	CEM IV/B²	45–64	For grouts and low-heat applications
Blastfurnace cement	BS EN 197-1	CEM IIIA	35–64	All types of general construction. Where slag level above 40% gives protection against alkali silica reaction. Moderate low-heat properties
Blastfurnace cement	BS EN 197-1	CEM IIIB	20–34	Where sulfate resistance and/or low heat properties required
Portland limestone cement	BS EN 196-1	CEM II/A-L CEM II/A-LL	80–94	For all types of general construction except where resistance to sulfates is required

*As percentage of nucleus, i.e. excluding gypsum.

and skin contact with wet cement, mortar or concrete should be avoided. Serious skin burns requiring skin grafts have occurred where concrete has penetrated workers' boots or where workers have kneeled for a prolonged period in fresh concrete wearing fabric trousers.

In addition to the acute burns described above exposure to moist mortar or concrete over a period of time may result in irritant contact dermatitis. Some people are sensitive (or after several years' exposure will become sensitized) to the water-soluble chromate present in cement. In Scandinavian countries ferrous sulfate has been added to cement during manufacture to precipitate the soluble hexavalent chromium as the insoluble trivalent form. The maximum permitted level of water-soluble chromium in these countries is

2 ppm. Although the clinical evidence for the effectiveness of this measure is inconclusive there are moves in Europe to adopt this approach more widely, at least in bagged if not bulk cement. The addition of ferrous sulfate has an influence on concrete rheological properties and also concrete colour. In the UK, the manufacturers have decided to monitor the level of soluble chromium and minimize levels by careful selection of raw materials.

References

Bhatty, J.I. (1995) *Role of Minor Elements in Cement Manufacture and Use*. Portland Cement Association, Skokie, PCA RD 109.2T.

Blezard, R. (1998) History of calcareous cements. In Hewlett, P.C. (ed.), *Lea's Chemistry of Cement and Concrete*. Arnold, London, pp. 1–19.

Bogue, R.H. (1955) *The Chemistry of Portland Cement* (2nd edn). Reinhold, New York.

BS 8500: (2002) *Concrete – Complementary British Standard to BS EN 206-1*. Part 2: Complementary requirements for constituent materials, designed and prescribed concrete, production and conformity. BSI, London.

BS EN 196-2: (1995) Methods of testing cement. Part 2. Chemical analysis of cement.

Cembureau (1991) *Cement Standards of the World* (8th edn). Cembureau, Brussels.

Coole, M.C. (1988) Heat release characteristics of concrete containing ground granulated slag in simulated large pours. *Concrete Research* **40**(144), 152–158.

Kelham, S. (1996) The effect of cement composition and fineness on expansion associated with delayed ettringite formation. *Cement and Concrete Composites* **18**, 171–179.

Kelham, S. and Moir, G.K. (1992) In *9th International Congress on the Chemistry of Cement,* Vol. 5. NCCBM, Delhi, pp. 3–8.

Killoh, D.C. (1988) A comparison of conduction calorimetry and heat of solution methods for measurement of the heat of hydration of cement. *Advances in Cement Research* **1**(3), 180–186.

Macfadyen, J.D. (2000) Online chemical analysis for control. *World Cement* **13**, 8, 70–73.

Massazza F. (1988) Pozzolana and pozzolanic cements. In Hewlett, P.C. (ed.), *Lea's Chemistry of Cement and Concrete*. Arnold, London, pp. 471–631.

Moir, G.K. (1994) Minor additional constituents. In Dhir, K.D. and Jones, M.R. (eds). *Euro-Cements Impact of ENV 197 on Concrete Construction*, E&FN Spon, London, pp. 37–56.

Moir, G.K. (1999) Factors influencing the ratio of mortar to concrete strengths. In Dhir and Jones (eds), *Proceedings of Conference on Creating with Concrete, Volume, Modern Concrete Materials: Binders Additives & Admixtures*. Thomas Dyer, Dundee.

Moir, G.K. and Glasser, F.P. (1992) *Proceedings of the 9th International Congress on the Chemistry of Cement*, Vol. 1. NBC, New Delhi, pp. 125–152.

Moir, G.K. and Kelham, S.K. (1997) Developments in the manufacture and use of Portland limestone cements. *Proceedings of ACI International Conference*, Malaysia. SP 172-42, pp. 797–819.

Moranville-Regourd, M. (1988) Cements made from blastfurnace slag. In Hewlett, P.C. (ed.), *Lea's Chemistry of Cement and Concrete,* Arnold, London, pp. 633–674.

Odler, I. (1998) Hydration, setting and hardening of Portland cement. In Hewlett, P.C. (ed.), *Lea's Chemistry of Cement and Concrete*. Arnold, London, pp. 260–263.

Oesbaek, B. and Jons, E.S. (1980) The influence of the content and distribution of alkalis on the hydration properties of Portland cement. *Proceedings of 8th International Congress on the Chemistry of Cement*. Septima, Paris, pp. II 135–140.

Taylor, H.F.W. (1987) A method for predicting alkali concentration in cement pore solutions. *Advances in Cement Research,* **1**(1), 5–17.

Taylor, H.F.W. (1997) *Cement Chemistry*, Thomas Telford, London, pp. 113–225.

Taylor, H.F.W., Mohan, K. and Moir, G. (1985) Analytical study of pure and extended Portland cement pastes: I, Pure Portland cement pastes. *Journal of the American Ceramic Society* **68**(12), 680–685.

2

Calcium aluminate cements

Karen Scrivener

2.1 Introduction

Calcium aluminate cements (CACs) are the most important type of non-Portland or special cements. Even so, the volume used each year is only about one thousandth of that of Portland cement. As they are considerably more expensive (four to five times), it is therefore not economic to use them as a simple substitute for Portland cement. Instead their use is justified in cases where they bring special properties to a concrete or mortar, either as the main binder phase or as one component of a mixed binder phase.

Some of the properties that can be achieved through the use of CACs are:

- rapid hardening
- resistance to high temperatures and temperature changes
- resistance to chemical attack, particularly acids
- resistance to impact and abrasion.

The uses of calcium aluminate cements today are extremely diverse. The largest single use is in refractory concretes, which is described in a separate chapter. CACs are also extensively used in combination with other mineral binders (e.g. Portland cement, calcium sulfate, lime) and admixtures to produce a range of specialist mortars for applications such as repair, floor levelling, tile adhesives and grouts.

Uses in traditional concrete are generally found where severe conditions must be faced, such as extreme temperatures, exposure to aggressive chemicals, mechanical abrasion or impact, often in industrial situations and where these severe conditions are combined with a need for rapid return to service. Some typical applications are illustrated in Figure 2.1.

This chapter aims to give the reader a broad understanding of the properties and behaviour of calcium aluminate cements as relevant to their present-day applications in order to support appropriate and intelligent use of these materials. The aim has been for clarity, focusing on the essentials with a minimum of references. For further reading and

Figure 2.1 Some typical applications of CAC concrete. From top left:
- Bridge widening, rapid hardening
- Dam flushing gate, abrasion resistance
- Foundry floor, abrasion and thermal shock resistance
- Cyrogenic facility, thermal shock, freeze–thaw
- Sewer, resistance to bacteriogenic acid corrosion
- Fire training building, resistance to thermal shock

details the reader is referred to Scrivener and Capmas (1998), Concrete Society (1997) and Mangabhai and Glasser (2001).

2.1.1 Terminology

As this is often a matter of confusion a brief explanation is given of some of the cement chemistry terms used.

A *phase* is the name given to distinct materials with a specific combination of *elements* with a specific arrangement, or *crystal structure*. In all the phases found in cements, elements such as calcium, silicon, aluminium and iron are always present with a fixed quantity of oxygen. This leads to the use of a shorthand notation, where letters are used to represent the fixed combinations of elements and oxygen or oxides:

$C = CaO$ (*lime*)
$S = SiO_2$ (*silica*)
$A = Al_2O_3$ (*alumina*)
$F = Fe_2O_3$ (*ferric oxide*)
$f = FeO$ (*ferrous oxide*)
$\$ = SO_3$ (*generally referred to as sulfate*)
$H = H_2O$ (*water*)

Cements are made by heating the raw materials at high temperatures to give a *clinker*, which is a combination of several different phases. As there is no water (or hydrogen oxide) present in these phases they are sometimes referred to as *anhydrous* phases. After reaction with water, new phases form, in which water is chemically combined with the other oxides. These phases are known as *hydrates*.

Some cement phases are also commonly referred to by their mineral name these are given on first mention of the phase for completeness and as information for further reading.

2.2 Chemistry and mineralogy of CACs

The original motivation for the development of CACs was to find new cement chemistries that would be more resistant to sulfate attack than Portland cement. This led various research groups to develop cements based on calcium aluminates and eventually to the patenting of a viable method of large-scale production by Bied of the Lafarge Company in 1908. Following this, other special properties, such as rapid hardening, were quickly discovered and used for a range of applications. For instance, in the First World War these properties were used in battlefields to build tank defences. However, calcium aluminate cements were not produced commercially until the 1920s.

Calcium aluminate cements, like Portland cements, contain the oxides of calcium, silicon, aluminium and iron. However, their composition is quite distinct as can be seen in Figure 2.2. This shows the approximate zone of compositions of calcium aluminate cements, within the $CaO–SiO_2–Al_2O_3$ system along with the composition zones of Portland cement and of blastfurnace slag.

Due to the requirements for refractory concretes, CACs are produced with a wide

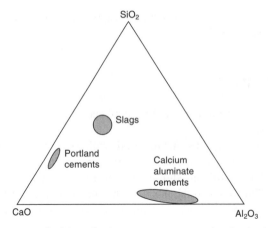

Figure 2.2 Composition range of calcium aluminate cements compared to Portland cements.

range of alumina (Al_2O_3) contents from around 40–80%. The higher alumina content grades are made from relatively pure raw materials (metallurgical alumina and lime), which make them very expensive. They are made in rotary kilns, similar to, but smaller than, typical Portland cement kilns. These grades are used almost exclusively for refractory concretes, which are dealt with in a separate chapter (although they have the same basic chemistry of hydration, etc.).

Here we are mainly concerned with the lower alumina grades (~40–60% Al_2O_3), which are generally made by complete melting of bauxite and limestone in a 'reverbatory' furnace*. The type of bauxite used determines the minor elements present and hence the colour of the cement. Cements made with red bauxite contain about 20% of iron oxides and have a brown, dark grey or black colour (for example, Ciment Fondu®[†]). Cements made with white bauxite contain little or no iron and are light grey to white. However, they still contain around 3–8% of silica (for example, Secar 51®[†]).

For these grades of CAC with lower alumina content the mineralogy (i.e. phases present) of the unreacted cement can be fairly complex, but the common feature of these, and all calcium aluminate cements, is the presence of monocalcium aluminate – CA or $CaAl_2O_4$. This is the principal reactive phase responsible for properties of the material and typically makes up from 40% to 60% of the cement. The other reactive phase is $C_{12}A_7$ (mayenite). This phase is important because it helps to trigger the setting process. However, its presence is strictly controlled by the manufacturers since too much would lead to early stiffening.

The other phases – C_2S (belite), C_2AS (gehlenite), spinel phase, ferrite solid solution, etc. (Touzo *et al.*, 2001) – contain silicon and/or iron and other minor elements in addition to calcium and aluminium. The combined quantities of these phases may make up over half of the cement. Therefore they are important from the point of view of

---

* In a reverbatory kiln, the raw materials in the form of large lumps slowly descend a vertical stack where heat exchange with the exhaust gases occurs. At the base of the stack is the hearth, where complete melting of the raw materials occurs. The molten product is tapped off continuously and solidifies in large ingots.
[†]Products of Lafarge Aluminates.

manufacture and correct control of the cement compositions, but these phases have much lower reactivity and play a negligible role in the initial hydration reactions, although they may partially react at later ages or higher temperatures. Figure 2.3 shows the microstructure

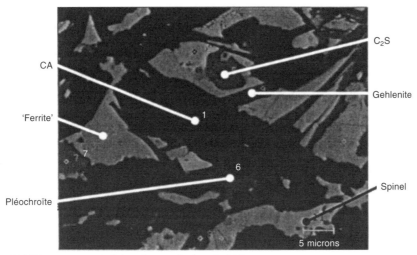

Figure 2.3 Microstructure of clinker of a standard grade CAC.

of a standard grade calcium aluminate cement, with the most important phases labelled. The size of the dark grey areas of monocalcium aluminate varies from grain to grain (Figure 2.4), due to the different cooling rate of the cement as it solidifies in the ingots,

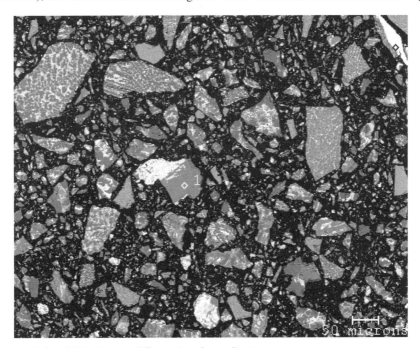

Figure 2.4 Grains of CAC, showing different size of crystallization.

after emerging as a melt from the furnace. The fine-textured regions come from the outer parts of the ingot, where the cooling rate is faster, and the coarser-textured regions from the centre, which cools more slowly. Despite the difference in the size of the areas, the chemical composition of the individual phases is very constant across the ingot. The areas of monocalcium aluminate are not individual grains, but are interconnected in three dimensions, so all of this phase is able to react without restriction by unreactive phases.

For standard grade CACs both the *chemical content of alumina* (Al_2O_3) and the content of the *phase, monocalcium aluminate* (CA) lie in the range 40–60%, but there is no direct equivalence between these two quantities. Most of the different *phases* contain alumina in differing amounts, and conversely CA, may contain small amounts of other elements (e.g. iron). The overall chemistry of a cement depends on the proportions and composition of the raw materials and can be measured either by traditional wet chemistry techniques or, more normally nowadays, by X-ray fluorescence (XRF). The phase composition or *mineralogy* is determined by this chemistry of the raw materials and thermodynamics. It may also be influenced by the parameters of cement production, e.g. furnace oxidizing conditions, cooling rate. However, the complexity of the mineralogy and the compositions of the individual phases mean that it is not possible to derive simplified relationships, equivalent to the Bogue calculation for Portland cements. Quantitative analysis of the phases present is now possible by X-ray diffraction combined with Rietveld analysis, but careful calibration and a good understanding of crystallography are necessary to obtain reliable results (Füllmann *et al.*, 1999).

2.2.1 Basics of hydration and conversion

The chemical process by which cements react with water to give a rigid solid is known as *hydration*. In terms of its physical process this is broadly the same for calcium aluminate cements as for Portland cements – namely anhydrous cement and water are replaced by hydrate phases, which have a larger solid volume and bridge the spaces between the cement grains. However, the products of calcium aluminate cement are *chemically* quite different from those of the calcium silicate phases found in Portland cements.

As with Portland cements, hydration occurs by a process of *dissolution* and *precipitation*. This means that the *anhydrous* phases in the grains of cement dissolve to give ions in solution, much as sugar or salt dissolve in water. However, unlike sugar or salt, the ions from the cement phases can recombine in different proportions with the ions of water. These new combinations come out of solution or *precipitate*, due to their lower solubility, as the hydrate phases.

When calcium aluminate cement (or monocalcium aluminate, CA) is placed in water calcium ions (Ca^{2+}) and aluminate ions ($Al(OH)_4^-$) dissolve in the water to give a solution. These can combine as several different types of hydrate generally known by their chemical formulas – CAH_{10}, C_2AH_8, C_3AH_6 and AH_3, plus poorly crystallized gel-like phases. In order to understand when and why the different hydrates form, and the practical consequences of this, it is useful to recall some notions of thermodynamics.

When different phases are mixed together, they will tend to react and recombine to give a new mixture of phases that has the lowest energy – this is the *stable* phase assemblage. However, sometimes it is quite difficult for the stable phases to form immediately due to the rearrangement of ions that needs to occur (*nucleation*). In this case it often happens that different phases form temporarily, which are *metastable*. They have a lower energy than the starting phase assemblage, which provides a driving force for their formation, but they have a higher energy than the stable phase assemblage. Thus a driving force remains for the meta-stable phases to in turn react to give the stable phase assemblage. As illustrated in Figure 2.5, this may be thought of as a 'step' in energy. The appearance of metastable phases is fairly common in other systems such as aluminium alloys of the type used in aircraft.

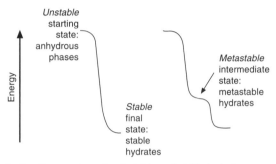

Figure 2.5 Scheme indicating the meaning of metastable and stable hydrates.

At all temperatures the stable hydrates of CA are C_3AH_6 (a form of hydrogarnet) and AH_3, (gibbsite):

$$3CA + 12H \Rightarrow C_3AH_6 + 2AH_3 \tag{2.1}$$

The crystal structure of C_3AH_6 is cubic so it usually has a morphology of compact equiaxed facetted crystals, (Figure 2.6), whereas AH_3 often is poorly crystalline and is deposited in formless masses. Figure 2.7 shows the microstructure of a 30-year-old concrete from a real structure (bridge at Frangey built 1969 with total w/c ~0.4). The C_3AH_6 is finely dispersed in a matrix of hydrated alumina.

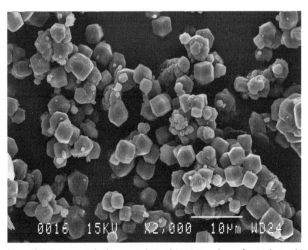

Figure 2.6 Morphology of C_3AH_6 crystal. The crystals in this picture have formed at a high water-to-cement ratio, so are quite large. Those formed in normal pastes during conversion are typically less than a micron in size.

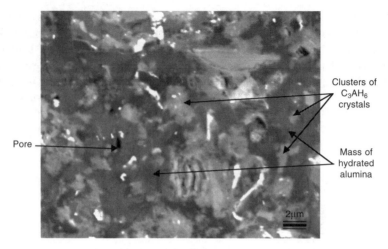

Clusters of
C_3AH_6
crystals

Pore

Mass of
hydrated
alumina

$2\mu m$

Figure 2.7 Microstructure of 30-year-old CAC concrete from structure (w/c ~0.4) – fully converted. A fine dispersion of C_3AH_6 (light grey) in a darker mass of hydrated alumina. There are very few, small isolated pores. The bright areas are unreacted iron rich anhydrous phases.

As these are the stable phases, other phases that may form initially will over time transform or *convert* to these phases.

At temperatures up to about 65°C the nucleation of C_3AH_6 is very slow. For temperatures up to about 27–35°C the first hydrate to form is usually CAH_{10}, while C_2AH_8 and AH_3 dominate initially between 35 and 65°C (Fryda *et al.*, 2001):

$$CA + 10H \Rightarrow CAH_{10} \tag{2.2}$$

$$2CA + 16H \Rightarrow C_2AH_8 + AH_3 \tag{2.3}$$

The subsequent conversion reactions to the stable phases are:

$$2CAH_{10} \Rightarrow C_2AH_8 + AH_3 + 9H \tag{2.4}$$

$$3C_2AH_8 \Rightarrow 2C_3AH_6 + AH_3 + 9H \tag{2.5}$$

All these reactions take place through solution, that is, the reacting phases dissolve and the product phases are precipitated from solution. Although thermodynamically it is possible for CAH_{10} to transform directly to C_3AH_6, C_2AH_8 will usually form as an intermediate phase due to ease of nucleation. Once the stable phases are present these will continue to form even if there is a fall in temperature.

From a practical point of view, the formation of these different hydrates is important because they have very different densities and contain different amounts of combined water (Table 2.1). Thus when the conversion of metastable to stable hydrates occurs:

1 There is a reduction in solid volume and so an increase in porosity and reduction in strength at an equivalent degree of hydration (see section 2.5)
2 Water is released which is available to hydrate any remaining anhydrous phases

The most straightforward analytical method to determine the type of hydrates present is differential thermal analysis. The metastable phases have decomposition peaks in the temperature range 100–200°C, while the stable phases decompose in the range 250–350°C (Figure 2.8).

Table 2.1 Density and combined water for calcium aluminate hydrates

Phase	Density (kg/m³)	Combined water (%)
CAH_{10}	1720	53
C_2AH_8	1750	40
C_3AH_6	2520	28
AH_3	2400	35

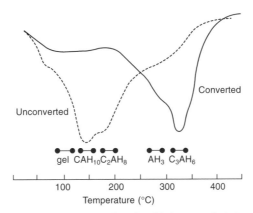

Figure 2.8 DTA traces of metastable (unconverted) and stable (converted) CAC pastes.

2.2.2 Impact of conversion

The impact of conversion is greatest at high water-to-cement ratios (≥ 0.7). In this case, if the temperature is kept low, there is sufficient water and space available for nearly all the reactive anhydrous phases to react to give meta-stable hydrates (CAH_{10} and C_2AH_8). Because these have a low *solid* density they fill most of the space originally occupied by water, leaving low porosity (Figure 2.9(a)). When conversion occurs the formation of the more dense stable hydrates (C_3AH_6 and AH_3) leads to a considerable decrease in solid volume, an increase in porosity and a consequent decrease in strength (Figure 2.9(b)). The volumetric relationships are shown schematically in Figure 2.9(c).

When the w/c ratio is kept low (i.e. <0.4), there is insufficient water and space available for all the cement to react to form metastable hydrates (Figure 2.10(a)). In this case the water released by conversion is available to react with more of the cement to give more hydrates. The net reduction in solid volume and increase in porosity is lessened and thus dense, low porosity microstructures are obtained after conversion (Figure 2.10(b)). The consequent decrease in strength is therefore less marked. Figure 2.10(c) shows the volumetric relationships in this case.

Although the water-to-cement ratio is the principal parameter controlling the properties of all concretes, this impact on microstructure before and after conversion emphasizes its special importance for calcium aluminate cement concretes.

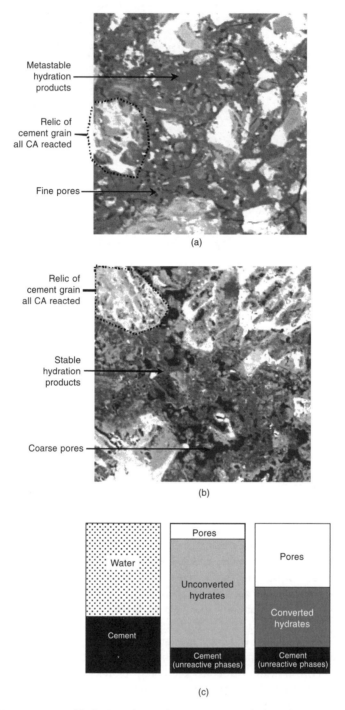

Metastable
hydration
products

Relic of
cement grain
all CA reacted

Fine pores

(a)

Relic of
cement grain
all CA reacted

Stable
hydration
products

Coarse pores

(b)

Pores

Water

Cement

Pores

Unconverted
hydrates

Cement
(unreactive phases)

Pores

Converted
hydrates

Cement
(unreactive phases)

(c)

Figure 2.9 Microstructures and hydration scheme of CAC concretes at *high* w/c (~0.7). When hydrated at 20°C ((a), top) a fairly dense microstructure can be produced, but this will convert to the coarse open microstructure as shown in (b) (middle) for a concrete hydrated at 70°C. The volumetric relationships are shown schematically in (c) (bottom).

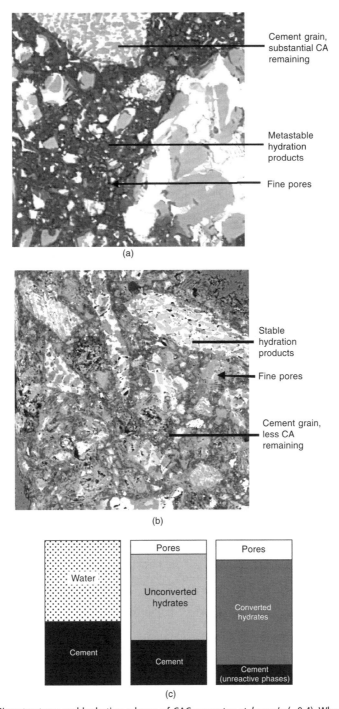

Figure 2.10 Microstructures and hydration scheme of CAC concretes at *low* w/c (~0.4). When hydrated at 20°C ((a), top) a dense microstructure is produced, after conversion the microstructure remains dense with the further hydration of unreacted cement ((b) middle). The volumetric relationships are shown schematically in (c) (bottom).

2.3 Properties of fresh CAC concrete – setting, workability, heat evolution

2.3.1 Setting

The setting times of calcium aluminate cements are similar to those of Portland cement, namely around 3–5 hours. For instance, the British standard for CAC, of the 'Fondu' type, requires that the initial set is not less than 2 hours and the final set not more than 8 hours, when measured on pastes at standard consistency (~24–28% water) by the Vicat needle method. The reason for this relatively long induction period before the onset of initial setting is the already mentioned delay in the nucleation of the hydrates.

The setting time of calcium aluminate phases becomes progressively longer with decreasing lime-to-alumina ratio. C_3A, the lime rich phase present in Portland cements, reacts rapidly with water, necessitating the addition of gypsum to Portland cement clinker to prevent flash setting. At the other extreme CA_2 reacts extremely slowly and CA_6 is inert at ambient temperatures. The setting time of pure CA is typically about 18 hours, but the shorter setting time seen in calcium aluminate cements is due to the disproportionate effect of the small quantities of lime-rich $C_{12}A_7$ phase present. Because of this strong effect of $C_{12}A_7$, variations in its amount can cause the setting time to change by hours and excessive amounts can cause early stiffening. The amount of $C_{12}A_7$ is therefore tightly controlled during production to avoid this variability in setting time and other properties that may also be affected as a consequence.

The setting time of CACs may also be affected unintentionally by contaminants or they may be intentionally adjusted by the use of additives. A number of contaminants, such as sugars, most acids, glycerine etc., have a serious retarding effect. Seawater also tends to retard setting. However, used intentionally, retarders such as citric acid or sodium citrate have beneficial effects on fluidity and working time.

The most common, effective accelerators for CACs are lithium salts (e.g. lithium carbonate) and these may be combined with retarding workability aids, such as sodium citrate, to improve workability while maintaining a normal setting time.

Accidental contamination with Portland cement will also accelerate the setting time of CAC (flash setting may occur if contamination is severe). However, this effect may be used in a controlled way, in complex formulated products.

As different hydrates form at different temperatures (as described above) the effect of temperature on setting time is more complex than in Portland cements. In pastes mixed in small quantities it is observed that the setting time increases with temperature up to about 27°C, before decreasing again as the temperature is further increased. However, in concretes mixed in normal quantities this effect is less pronounced due to a range of effects including self-heating and the heating due to the friction during mixing of the aggregates present in the concrete (Figure 2.11).

2.3.2 Workability

In the absence of admixtures the intrinsic ease of placing CAC concretes is similar to that of Portland concretes at the same water-to-cement ratio. CAC concretes appear 'dryer', but the temptation to add water must be avoided as they flow well under vibration.

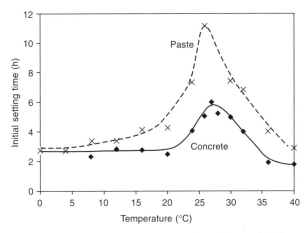

Figure 2.11 Effect of temperature on setting times. From Cottin and George (1982).

Classic plasticizers such as ligno-sulphonate and sodium citrate work reasonably well with CACs, but there may be a significant retarding effect. Superplasticizers based on naphthalene and melamine sulphonate are not very effective. Until recently, the best way of obtaining reasonable workability at the recommended low w/c ratios was to use a high cement content, typically more than 400 kg/m^3.

For the same reason it is important to have a well-graded aggregate, particularly sufficient fine sand, to minimize the volume between the aggregates that the cement paste needs to fill.

The modern generation of superplasticizers (e.g. PCP type) are highly effective with CACs, but the dosage must be carefully controlled to avoid excessive retardation. At a free w/c of around 0.35 it is now possible to obtain flowing concrete for more than 2 hours. These concretes still have strengths of around 30 MPa at 8 hours even at ambient temperatures of only 5°C (Fryda and Scrivener, 2002).

2.3.3 Rate of reaction and heat evolution

At the end of the induction period rapid reaction occurs. In Portland cement the main hydration product is deposited around the hydrating cement grains, so that after the first few hours the rate of reaction is progressively slowed down. In CACs, in contrast, the hydration products are deposited throughout space (Figure 2.12). The rate of reaction only slows due to one of the reactants (water or cement) being used up or to lack of space for the hydrates to precipitate. This pattern of reaction leads to the rapid hardening of CAC concrete and it is quite easy to obtain strengths of more than 20 MPa at 6 hours and over 40 MPa at 24 hours. The other consequence of the rapid reaction is that the heat of hydration is evolved over a relatively short time. As the heat generation is rapid relative to the rate of heat dissipation, self-heating may be considerable in sections more than about 100 mm in size. It is quite usual to have maximum temperature in the range 50–80°C or even higher. Figure 2.13 shows the temperature profile in a beam of concrete. Such self-heating effects are very important for the strength development on CAC concrete, as discussed below (section 2.5).

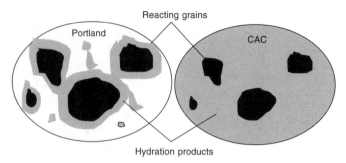

Figure 2.12 Schematic representation of the deposition of hydration products for the hydration of Portland and calcium aluminate cement.

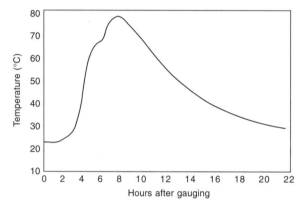

Figure 2.13 Temperature rise in beam 35 × 180 cm of CAC concrete. Formwork was removed after 5 hrs then beam sprayed with water for 20 hours. Adapted from French Standard (1991).

Despite these high temperature rises, CAC concretes do not appear to be overly susceptible to thermal cracking. This may be due to the fact that creep is facilitated by the conversion reaction and relaxes thermally induced strains.

An important consequence of the rapid temperature rise is the need to keep the concrete wet to prevent the drying and dehydration of the surface. For large sections it is recommended to remove the formwork as early as possible (around 6 hours) and spray the surface with water.

One advantage of this self-heating is that concreting can be continued in periods of very cold weather and indeed at sub-zero temperatures, provided the concrete is not allowed to freeze before hydration commences.

2.3.4 Shrinkage

Intrinsically the shrinkage of calcium aluminate cement is roughly similar to that of Portland cement at the same degree of hydration. However, as hydration occurs much more rapidly this shrinkage occurs over much shorter period of time and can lead to problems with early age cracking (Figure 2.14).

In order to minimize these effects the same rules of concrete technology that are known for Portland cements apply, namely:

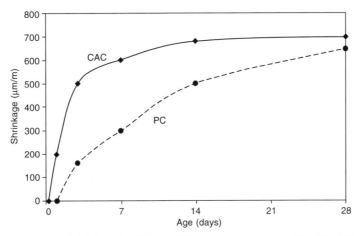

Figure 2.14 Comparison of shrinkage for CAC and OPC concrete measured in 40 × 40 × 160 prisms at 20°C, 50% RH. Data from French Standard (1991).

- prolonged moist curing
- good aggregate grading to minimize paste volume
- limitation of bay sizes (to around 3 m in unreinforced slabs)
- use of polymeric fibres in thin sections.

2.4 Strength development

Due to its hydration, with the successive formation of metastable and stable hydrates, the strength development of CAC concretes is more complicated than that of Portland cement concrete. However, there are a few straightforward facts to remember:

1 Conversion is a thermodynamically inevitable process, so the converted strength should be used for design.
2 Once conversion has occurred, stable hydrates are present and therefore the strength is also stable and is likely to increase due to further hydration of any remaining unreacted cement.
3 As with Portland cement concretes, and indeed any other porous material, the major factor related to strength is the porosity (or lack of it) and hence the water/cement ratio and the degree of reaction.
4 At equivalent water to cement ratios the converted strength of CAC concrete is slightly lower than that of standard Portland cement concretes.

For small elements maintained at temperaures around 20°C or lower, strength development follows the pattern shown in curves A and B in Figure 2.15 – this is the pattern of strength development often associated with the conversion of calcium aluminate cements showing substantial loss of strength. In fact it is rare to find such extreme cases in practice due to the substantial self-heating effects already discussed. Even in such extreme cases it is important to realize that the strength decreases to a stable minimum depending on the water-to-cement ratio and does not continue decreasing to zero.

In such cases the initial hydration reactions during setting and hardening produce a

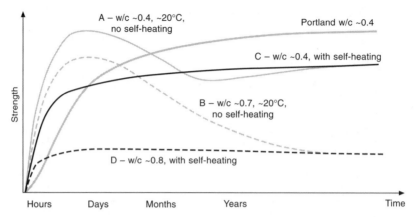

Figure 2.15 Schematic strength development of CAC concrete at high and low w/c, under isothermal curing at 20°C and under conditions of self-heating, compared with Portland cement concrete.

microstructure of metastable hydrates that combine a large amount of water and have a relatively high specific volume (low density). They form throughout space (as explained previously) and therefore create a microstructure wherein the porosity is very low (Figure 2.9(a)). Consequently, it is possible to obtain quite high strengths even at high water-to-cement ratios.

However, as the metastable hydrates convert to the stable higher density phases with less water in their structure there is a decrease in volume of hydrates and so an increase in porosity and a decrease in strength. For concrete maintained at ambient temperatures of 10–20°C this process may take ten years or more (still a comparatively short time in the life of most concrete).

Once conversion has occurred the strength is stable, but if the original water-to-cement ratio is low, the strength may increase after the minimum as water released by hydration reacts with remaining unreacted cement.

In practice, for concrete sections more than a few tens of centimetres in size, the heat generated by hydration will cause the temperature of the concrete to rise, typically to 50–80°C or more. In these conditions the stable hydrates are quickly formed since higher temperatures accelerate the rate of conversion as shown in Figure 2.16. For isothermal curing at 38°C the minimum strength occurs at 5 days. At 50°C the minimum is reached by around 24 hours and for temperatures greater than 50°C conversion occurs within a few hours.

In the case of many civil engineering concretes the size of the pour is sufficient for stable hydrates to be formed during the initial phase of curing and consequently there is no regression in strength, which will continuously increase with increasing hydration. This situation is illustrated in curves C and D of Figure 2.15.

2.4.1 Importance of water-to-cement ratio and control of concrete quality

As previously stressed, the control of water to cement (w/c) ratio is particularly important in CAC concretes due to its impact on the stable converted strength. The converted

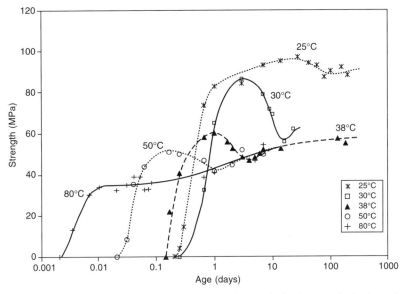

Figure 2.16 Impact of temperature of rate of conversion and strength development for isothermal curing. Adapted from French *et al.* (1971).

strength of CAC concrete decreases as w/c increases (Table 2.2). Historically total rather than free w/c ratios were used in studies of CAC concrete. Approximate equivalent free w/c are given in the table (assuming an aggregate absorption of around 1.2%). However, as CAC reacts much faster than Portland cement, there is less time for water to be absorbed by the aggregates and the effective w/c may be higher.

Table 2.2 Variation in converted strength with water-to-cement ratio (from George, 1976)

Total w/c	Approx free w/c	Minimum converted strength
0.33	0.28	53
0.4	0.35	47
0.5	0.44	37
0.67	0.58	27

Based upon considerable experience and laboratory testing, recommendations for 'total w/c' ratios were established. These manufacturer and industry recommendations are that the maximum total w/c ratio should be no greater than 0.4 (equivalent to free w/c ratio of ~0.33–0.36). With this maximum w/c the converted CAC concrete has good residual strength and low permeability.

However, the consequences on stable strength of using a high w/c ratio may be masked if the concrete strength is controlled at early ages on small samples cured at ambient temperature – because such a concrete would contain metastable hydrates. These samples may exhibit relatively high strengths, but this is misleading since they will eventually undergo a significant strength loss with time due to conversion.

The control of the strength of calcium aluminate cement as delivered on-site should not be made by casting cubes or cylinder in metal or cardboard moulds. For the reasons already discussed, control specimens cured at or near 20°C will have high unconverted strengths after one day, which are not representative of the ultimate strength of the *in-situ* concrete. Being insensitive to the high water-to-cement ratios used, such samples would give a false representation of the long-term strength.

A well-proven procedure that gives an accurate indication of the long-term design strength is to place control specimens immediately at a temperature of 38°C at 100% R.H. (or sealed). The minimum, converted, stable strength will occur at 5 days.

Alternatively insulated moulds can be used, in which the heat of hydration is retained, which means that the temperature profile of the control specimens is similar to that of the concrete *in situ*. Such test methods can give a useful forecast of the minimum strength but should not be relied upon without reliable quality assurance of the water addition for all batches on-site.

Site-verification acceptance of batched concrete is an issue for most concrete irrespective of cement type. This is most often undertaken by a combination of batching/mixing QA procedures and acceptance testing. For CAC concretes, the assurance of conformity with the maximum w/c is particularly important for long-term performance. Where CAC concrete is being used to shorten the time before reintroducing critical elements into service, the time required for conventional physical testing may be inappropriate and other routes may need to be considered. However, CAC concrete can be used with confidence if appropriate assurance of conformity can be properly achieved.

2.4.2 Other factors affecting conversion and strength development

As conversion is a chemical reaction occurring through solution, the presence of moisture is essential for it to occur. However, it is unlikely and unreasonable to assume that any concrete could be maintained in sufficiently dry conditions for conversion to be avoided. This is illustrated in Figure 2.17, which shows little difference in the strengths of concrete

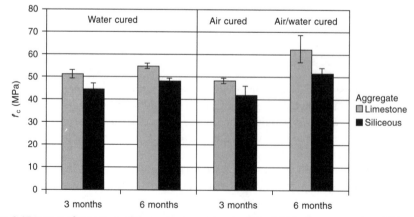

Figure 2.17 Impact of storage conditions on compressive strength of CAC concrete cured at 40°C. From Lamour *et al.* (2001).

at w/c = 0.4 stored in either air or water at 40°C. Subsequent storage of the air cured specimens in water, led to an increase in strength, probably through hydration of remaining anhydrous material.

In recent years several proposals have been made for systems in which conversion is inhibited (e.g. Majumdar and Singh, 1992; Osborne, 1994; Ding *et al.*, 1996). These rely on additions of material containing reactive silica – slag, silica fume, metakaolin, etc. For temperatures below about 50°C, the stable phase, strätlingite (C_2ASH_8), forms instead of C_3AH_6 by the reaction of the metastable hydrates with silica and, if the amount of addition is high enough, little or no strength regression due to conversion occurs. However, very high additions of these materials impair the early strength development. Also, if the concrete experiences temperatures above 50°C, as may occur through self-heating, C_3AH_6 will still be formed.

Another factor with an important impact on strength is the kind of aggregate used. Considerably higher strengths are obtained with calcareous aggregates, compared to silicons aggregates.

2.5 Other engineering properties

There are relatively few studies of other engineering properties at w/c ratios around the recommended 0.4. A recent study showed that the tensile strength and toughness were roughly the same as for Portland cement concretes (of comparable compressive strength) (Lamour *et al.*, 2001) and that these properties tended to be slightly less affected by conversion than the compressive strength. As for Portland concrete, the aggregate has a major impact on the elastic modulus, which otherwise is similar to Portland cement concretes of a similar strength. Poisson's ratio varied between 0.15 and 0.25, again similar to Portland concretes.

Relatively few studies of creep are reported, especially in the w/c range now used. However, it appears that the rate of creep depends on the applied stress/strength ratio in a manner similar to that of Portland cement.

2.6 Supplementary cementing materials

A major trend in Portland concrete technology is the increasing use of supplementary cementing materials – fly ash, slag, silica fume, fine limestone, metakaolin, etc. – to improve durability and reduce clinker consumption. As mentioned above, the presence of reactive silica favours the formation of strätlingite. In moderate amounts this may have a positive effect on the minimum, stable strength although the high additions necessary to completely eliminate strength regression lead to poor early strengths.

In other regards the impact of such additions on calcium aluminate cement has not been extensively studied. There is some evidence that they may contribute to improvements in rheology and allow the formulation of self-placing concretes.

2.7 Durability/resistance to degradation

As with all concretes, the primary factor determining durability is a good-quality concrete with low porosity. This implies the use of low water-to-cement ratios. Unconverted CAC concretes have exceptionally low permeability. Although the permeability increases with conversion reasonably low levels can still be achieved in converted concrete made with total w/c ratios at or below 0.4. Studies of old structures indicate that CAC concrete forms a dense surface layer, which may contribute to the good performance for over 70 years seen in these massive structures even with total water-to-cement ratios around 0.6 (Scrivener *et al.*, 1997a).

In addition to the specific forms of degradation discussed below, it should also be noted that no cases of alkali aggregate reaction have ever been observed with CAC concrete, due to its lower alkalinity (around pH 12.4 in comparison to >13 for Portland; Mácias *et al.*, 1996).

2.7.1 Reinforcement corrosion

Corrosion of reinforcement is well known to be the major cause of durability problems in Portland cement concrete. This is due to impairment of the passive oxide film, which normally protects reinforcing steel in the alkaline environment of the concrete, due to either reduction in this alkalinity through carbonation or the penetration of chloride ions.

Calcium aluminate cement concrete has a slightly lower alkalinity than Portland concrete, but the pH is still more than high enough to ensure the formation of a passive film on steel and so protection against corrosion under normal circumstances. The equilibrium pH for the metastable phase assemblage is 12.13 while that for the stable phase assemblage is only slightly different at 11.97 (Damidot, 1998). In real concretes the pH levels are higher (~12.5), due to the small amounts of alkali metal oxides impurities present, which remain in the pore solution (Mácias *et al.*, 1996). Conversion does not therefore change the intrinsic capacity of CAC to protect reinforcement.

As with Portland cement, both carbonation and penetration of chloride ions may lead to the breakdown of the passive layer, and to active corrosion occurring. Again, as with PC concrete, these processes are controlled by transport of CO_2 or chloride through the concrete cover, the rate of which is controlled by the resistance to permeability or diffusion of the concrete cover, which is in turn primarily a function of its porosity and hence the w/c ratio used and the degree of curing. The conversion in CAC concretes then becomes important in that it can lead to a highly porous concrete if a high w/c ratio is used.

Field evidence demonstrates that CAC concrete is capable of providing good protection to reinforcement over a sustained period of time. One of the largest structures ever built with CAC concrete was part of the ocean harbour in Halifax, Nova Scotia, Canada. Some 6000 tonnes of CAC was used for its construction between 1930 and 1932. Cores taken from the structure in the early 1990s showed that the reinforcing steel was in good condition and that penetration of chloride and sulfate was generally low. The formation of a dense surface layer at the surface of the concrete is thought to have contributed to its good durability.

2.7.2 Sulfate attack

CACs were originally developed for improved sulfate resistance and early CAC concrete was successfully used in tunnelling works through gypsum deposits. CAC concretes have consistently performed extremely well in the field, both in structures and in trials, provided the water-to-cement ratio is kept to the recommended levels. More variable results have been obtained in laboratory studies, but in such studies the pH of the storage solution is often not adjusted to keep it at the relatively low levels found in the field. Removing the surface layer of cubes also had a detrimental effect on their resistance to sulfate solutions. For more detail see Scrivener and Capmas (1998).

2.7.3 Freeze–thaw damage

Little laboratory data exists on the freeze–thaw resistance of calcium aluminate cement. Those studies that have been made indicate that CAC concretes show good freeze–thaw resistance when the porosity is low (concretes with a total w/c ratio less than 0.4). Field performance, especially where CAC concrete has been used in cold climates, have not identified any particular problem of freeze–thaw resistance. CAC concretes can be air entrained.

Studies of CAC concretes made with special calcium aluminate aggregate (Alag^{TM}) have shown that this type of concrete has exceptionally good freeze–thaw resistance. A very particular application where very good performance is achieved is in areas where there is spillage of liquid gasses ($T \sim -170°C$, Figure 2.1).

2.7.4 Alkaline hydrolysis

A form of degradation that is particular to calcium aluminate cements is alkaline hydrolysis. This is a very rare form of degradation that only occurs in concretes of very high w/c (> 0.8). In addition, an external source of alkalis and moist conditions are required. The mechanism is believed to be the leaching of alumina in highly alkaline conditions. Many explanations also focus on the role of carbonation. The role of CO_2 in exacerbating degradation by alkalis is probably that carbonation leads to a breakdown of the calcium aluminate hydrates into calcium carbonate and alumina, thus making the alumina more susceptible to alkaline leaching in highly porous concretes.

Microstructural examination reveals that concrete suffering from this form of degradation is extremely porous, leading to dramatic strength loss. The source of alkali can be from the aggregates or external, for example water percolating through Portland concrete or fibre boards or alkaline cleaning fluids. Laboratory studies followed for nearly 30 years confirm the critical role of w/c. In two series of specimens – one with alkali-containing aggregate and the other where water was absorbed through Portland concrete – degradation only occurred for the highest water-to-cement ratios – 1.0 and 1.2.

Long-term field studies have shown no problems in situations where CAC concrete (of normal w/c) adjoins Portland concrete (Deloye et al., 1996).

2.8 Structural collapses associated with CAC concrete

Many civil engineers will have heard of the structural collapses involving CAC concrete that occurred in the UK in the 1970s. These are treated in detail in a recent Concrete

Society (1997) report. The origin of these events was the use of CAC concrete in pre-cast, pre-stressed beams, which occurred in the post-war construction boom. At the time, conversion was known about, but the advice was conflicting. Rather than conversion being regarded as inevitable, it was considered to be negligibly slow if warm moist environments were avoided (Code of Practice 114 (1957) and annex (1965).

The importance of low water-to-cement ratios was also known, but recommendations (Code of Practice annex 114, 1965) to keep the w/c to 0.5 or less conflicted with the requirement of a cement content *less* than 400 kg/m^3, to limit the effects of self-heating. Concrete containing less than 400 kg/m^3 of cement would be most difficult to place at w/c < 0.5.

Nevertheless, loss of strength due to conversion was not the sole or primary cause of failure. In two out of the three incidents design problems were the primary cause – lack of adequate bearing area between the beams and the wall supports, which may have been exacerbated by thermal movement of the beams, which were not adequately linked to the walls. In one of these cases the original beams were tested to destruction and found to have adequate strength for the design, the remaining beams being re-used in the rebuilding. In the third case, the concrete was found to have degraded due to external sulfate attack from adjacent plaster. This attack was facilitated by the high porosity of the concrete due to the use of a high w/c.

In the wake of these events, occurring shortly after a major structural collapse of a tower block built with OPC concrete (Ronan Point), measures were taken which effectively prevented the use of CAC concrete in structural applications. The overall number of buildings containing CAC concrete built during this period which continue in service is believed to be around 30 000–50 000. After the collapses some 1022 of these were inspected, of which 38 were identified as having problems, but in only one of these was this attributed to loss of strength due to conversion.

Pre-cast, pre-stressed CAC concrete beams from a similar era were also implicated in problems in Spanish apartment buildings in the 1990s. The worst of these was the collapse of the 'Tora de la Peira' apartment block, in which beams made with both CAC and Portland concrete failed (George and Montgomery, 1992). In these cases the CAC concrete was degraded by alkaline hydrolysis due to poor-quality aggregates and highly porous concrete.

These unfortunate episodes illustrate the importance of understanding the materials behaviour of CAC concrete. The use of CAC concrete in small sections, where conversion does not occur rapidly due to self-heating, raises the difficulty of ensuring that the water-to-cement ratio is strictly controlled. Nowadays, steam-curing techniques mean that pre-cast beams with high early strength can be satisfactorily produced in Portland concrete at much lower cost than with CAC. It is unlikely that primary 'bulk' structural use of CAC concrete would be viable or justified.

The conclusion of the Concrete Society working party on CAC in construction which reported in 1997 (Concrete Society, 1997) was that there were situations in which the CAC concrete could be used safely and advantageously in construction, provided that its long-term converted strength is taken into account for design purposes. This has led to a change in the guideline to the UK building regulations, allowing use of materials whose properties change over time, provided that their residual properties are predictable and adequate for the design. However, there are still no BS codes of practice for the structural use of CAC concrete.

2.9 Modern uses of CAC concrete

The modern uses of CACs are based on areas where their special properties provide a major advantage over other cements.

2.9.1 Resistance to acids/sewage networks

The excellent resistance to acids, particularly those produced by bacteria, leads to use in sewage networks (Figure 2.1) (Scrivener *et al.*, 1999; Letourneux and Scrivener, 1999). In sewers, bacteria in the effluent produce hydrogen sulphide gas (the smell of rotten eggs), which is in turn used as a nutrient by another group of bacteria, resulting in the production of sulphuric acid. This can lead to very severe degradation of concrete, particularly in the crown and along the water-line of pipes. The growth of bacteria is most favourable at 30°C so such attack has become a notable problem in countries with hot climates.

The extensive positive field experience with CAC pipes is supported by the results of a 12-year field trial in South Africa (Goyns, 2001). The images shown in Figure 2.18 clearly illustrate the superior performance of CAC concrete. Reinforced pipes of 60 mm of different materials were placed along the same length of sewer. After 12 years the Portland concrete with siliceous aggregate has been completely eroded through the 60 mm thickness so that the earth is visible (Figure 2.18(a)). A section made with Portland cement and dolomitic limestone showed only marginally better performance.

In contrast, the section made with CAC concrete (with siliceous aggregate) was in a much better condition (Figure 2.18(b)) with a maximum depth of attack of around 10–15 mm only at the water-line. The reasons for this improved performance are:

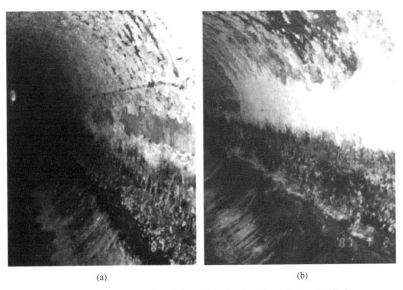

(a)   (b)

Figure 2.18 Sewage pipes after 12 years in a field trial in South Africa. The Portland/siliceous aggregate section (a) has completely eroded through its 60 mm thickness in places. The CAC/siliceous aggregate section (b) shows maximum erosion of 10–15 mm only at the water-line.

- The presence of hydrated alumina, which is insoluble in acid down to pH 4 and has a high neutralization capacity at lower pHs
- The inhibition of bacterial activity, which limits the surface pH to around pH 3.5 in contrast to pHs around 1 which develop on Portland concrete.

Studies have shown that even better resistance can be provided by using CAC in conjunction with a sacrificial dolomitic aggregate or (best of all) a synthetic aggregate (Alag™).

2.9.2 Resistance to abrasion and impact

CAC concretes show very good resistance to abrasion and impact when made with suitable hard aggregates and particularly with Alag™, which is a synthetic aggregate made from CAC clinker. This property has led to their use in a number of applications including ore passes in South Africa, Canada and Australia; in hydraulic dams in Switzerland, France and Peru and in sections of roads subject to heavy wear such as motorway toll booths (Scrivener *et al.*, 1999).

For Portland concrete it is generally found that abrasion resistance is a simply related to compressive strength. CAC concrete shows superior resistance to abrasion compared to Portland concretes of similar strength (Figure 2.19) (Saucier *et al.*, 2001). This

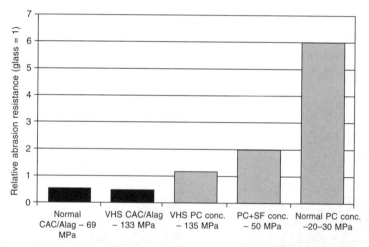

Figure 2.19 Relative abrasion resistance of CAC/Alag and Portland cement concretes. The very high strength (VHS) concretes have similar strength but very different abrasion resistances. Data from Saucier *et al.* (2001).

improvement is believed to be due to the improvement in the quality of the interfacial transition zone (ITZ) between the cement paste and aggregate. Normally in Portland concretes the ITZ is more porous, due to the difficulty of packing the cement grains in this region ('wall effect'); as the CAC hydrates form throughout space they infill this region better as illustrated in Figure 2.20. Partial reaction at the surface of the synthetic aggregate also improves the bond.

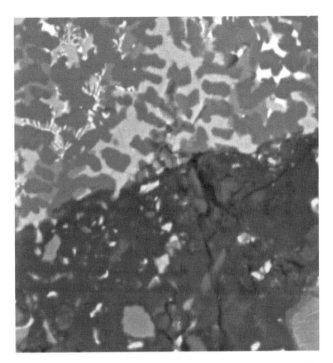

Figure 2.20 Interfacial transition zone (ITZ) between CAC and Alag aggregate. Due to formation of hydrates throughout space the ITZ has low porosity, unlike in Portland cement concretes.

2.9.3 Rapid strength development

CAC concrete can provide unique technical advantages, when very rapid strength development is needed, for example industrial applications where a rapid return to service is important. Another example is the pre-cut tunnelling method where a CAC/slag blend is used to provide a temporary concrete lining, prior to excavation of the main tunnel bore (Defargues and Newton, 2000). The concrete is applied as shotcrete and develops a strength of over 8 MPa in 4 hours, which allows the next tunnel segment to be excavated. A lining of conventional concrete is installed in the final tunnel.

The rapid strength development of CAC concrete is particularly advantageous in cold climates and it has been widely used in the Arctic – for example, for grouting cable stays. Even at 0°C CAC concrete can develop strengths of 20–30 MPa at 16 hours (Scrivener and Capmas, 1998).

2.9.4 Thermal resistance

CAC concrete is very resistant to heating and thermal shock. This property is explained in more detail in a later chapter, which discusses heat-resisting and refractory concretes. However, this property is also used to provide resistant floors in (for example) foundries (Figure 2.1). Another use drawing on this property is in buildings for fire training, which can withstand repeated lighting and extinguishing of fires (Figure 2.1).

2.10 Use of CACs in mixed binder systems

An increasingly important use of calcium aluminate cements today is as a component of mixed binder systems for specialist applications. The ingredients are usually blended dry and sold as mortars ready to mix with water. The applications are typically non-structural finishing operations in buildings, such as floor levellers, tile adhesives or fixing mortars. Another application is rapid-hardening repair mortars.

Such mixed binders are usually based on combinations of three reacting ingredients – calcium aluminate cements, Portland cement and calcium sulfate. The calcium sulfate may be in the form of gypsum, anhydrite or plaster (calcium sulfate hemi-hydrate). Fillers such as limestone and supplementary cementitious materials such as slag may also be present. In addition these mixes usually contain several admixtures in dry form (accelerators, retarders, fluidifiers, plasticizers, thickeners). Fine siliceous sands (and sometimes fine aggregate) are also normally used in these special mortars as they would be in conventional mortars.

Figure 2.21 shows the common zones of formulation of the CAC, PC and C$ components. **Zone 1** concerns blends based on binary mixtures of Portland and calcium aluminate cement (with more of the former) to which smaller amounts of calcium sulfate may be added. These mixtures are rapid setting and hardening.

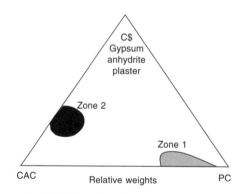

Figure 2.21 Zones of formulation in 'Building Chemistry'.

It has long been known that additions of CAC to Portland cement can induce flash setting. The basis of this behaviour is that the added calcium aluminates combine with the calcium sulfate in the Portland cement added during grinding. This disrupts the normal set control process of the Portland cement, in which calcium sulfate reacts with the rapidly reacting C_3A phase of the Portland clinker, producing products which coat the surface of the C_3A and block its further reaction for several hours. The relative rates of dissolution of the various forms of calcium sulfate (anhydrite, hemi-hydrate and gypsum) are different, so the amount of CAC that needs to be added to give flash setting varies with the amount and type of calcium sulfate in the Portland cement.

Such simple binary mixtures may show rather poor long-term strength development. This can be improved by the addition of admixtures or gypsum. Further discussion of the mechanisms occurring in such system can be found in Amathieu et al. (2001). As an example a correctly formulated mixture in this zone can have a working time of 2 hours and shows strength of around 20 MPa after 6 hours.

In such systems the hydrated microstructure is similar to that of Portland cement, with calcium silicate hydrate being the dominant phase. The quantity of calcium hydroxide will be decreased and the quantities of ettringite and/or monosulfate increased according to the relative amounts of CAC and calcium sulfate contained in the blend. For the amounts of CAC typically used, the metastable calcium aluminate hydrates are not formed so there is no question of conversion.

The mixtures in **Zone 2** are composed of roughly equi-molar amounts of monocalcium aluminate (CA) from CAC and calcium sulfate, with minor amounts of Portland cement. The predominant hydration reaction in this zone is:

$$3CA + 3C\$ \cdot H_x + (38\text{-}x)H \Rightarrow C_3A \cdot (C\$)_3 \cdot 32H + 2AH_3 \tag{2.6}$$
$$\text{ettringite}$$

In equation (2.6) the formula AH_3 for gibbsite is used for convenience, but this phase is often poorly crystalline. Microstructural examination shows crystals of ettringite dispersed in a dense alumina-rich matrix. Both these phases are stable phases and will not spontaneously convert to other phases.

Such mixtures can be controlled to be fluid for a few hours to allow placing and then harden rapidly. However the main difference, between the mixtures of this zone and those of zone 1 is their capacity to combine water in the hydrates and so to rapidly reduce the relative humidity. This leads to their major application as floor-levelling compounds (Figure 2.22), where the rapid reduction of relative humidity (commonly referred to as drying) allows them to be overlaid with carpet or floor tiles with a minimum of delay (as little as 8 hours). Table 2.3 shows the percentage of combined water by weight in various hydrate phases.

Figure 2.22 Application of floor-levelling compound.

Table 2.3 Percentage of water combined in various hydrates

Phase	Formula	% H_2O
ettringite	$3CaO \cdot Al_2O_3 \cdot 3CaSO_4 \cdot 32H_2O$	46
CAH10	$CaO \cdot Al_2O_3 \cdot 10H_2O$	53
monosulfate	$3CAO \cdot Al_2O_3 \cdot CaSO_4 \cdot 12H_2O$	35
gibbsite	$Al(OH)_3$	35
C_2AH_8	$2CaO \cdot Al_2O_3 \cdot 8H_2O$	40
C_3AH_6	$3CaO \cdot Al_2O_3 \cdot 6H_2O$	28
CSH	$1.7CaO \cdot SiO_2 \cdot 4H_2O$	30
Portlandite	$Ca(OH)_2$	24

2.10.1 Stability of ettringite-containing systems

As discussed, ettringite is an important component in all these mixed systems and questions are posed as to its stability, regarding temperature, humidity, and carbonation. On the first two issues, recent work published by Zhou and Glasser (2001) confirms that ettringite has broadly similar stability to temperature and humidity as C–S–H the principal hydrate found in hydrated Portland cement. It is important to realize that ettringite occurs as part of a system of hydrates, existing as a rigid, if porous, solid and not as an isolated powder. Thermogravimetric analysis shows that the decomposition temperatures of ettringite and C–S–H are very similar, around 110–130°C.

On the question of carbonation, analogies may also be drawn with conventional Portland systems. Finely powdered ettringite will carbonate rapidly, as will finely powdered C–S–H, but when either are part of a hydrate system the kinetics of the reaction are controlled by the ingress of CO_2 into a porous solid. The carbonation of ettringite eventually leads to calcite, gypsum and alumina:

$$3CaO \cdot Al_2O_3 \cdot 3CaSO_4 \cdot 32H_2O + 3CO_2 \Rightarrow 3CaCO_3 + 3CaSO_4 \cdot 2H_2O \qquad (2.7)$$

$$+ 2Al(OH)_3 + 9H_2O$$

This reaction leads to the release of water, which helps to maintain a high relative humidity inside the solid, which impedes the ingress of CO_2 gas. However, the loss of water corresponds to a decrease in the solid volume and an increase in porosity. Therefore, there will be some loss of strength in materials containing ettringite that become completely carbonated. For formulations in Zone 1, such strength losses are negligible and will be compensated for by the continuing increase in strength due to hydration. For formulations in zone 2 the loss in strength is more significant, but not catastrophic. Laboratory studies (Scrivener *et al.*, 1997b) of accelerated carbonation of a composition which had a one-day strength of ~30 MPa showed that the samples stored at 100% CO_2 and 60% R.H. had stable strength of 35 MPa at 3 months (corresponding to complete carbonation) as opposed to a strength of 45 MPa for the reference sample stored at 60% R.H. with atmospheric CO_2.

2.10.2 Dimensional stability of ettringite-containing systems and shrinkage compensation

Another issue related to ettringite is that of dimensional stability. By formulation it is also possible to engineer the initial hydration reaction to produce controlled expansion and so compensate for the normally observed drying shrinkage of Portland cement. This is similar to the actions of shrinkage compensation in type K cements.

It is well known that the formation of ettringite is also related to undesirable expansion during external sulfate attack and in so-called internal sulfate attack or delayed ettringite formation (DEF) which may occur after exposure to temperatures >70°C during curing. However, the relationship between ettringite formation and expansion is extremely complex. For example, after exposure to temperatures >70°C during curing delayed ettringite forms in Portland cement mortars or concretes, irrespective of whether expansion occurs or not. Extensive studies have shown that the crystallization of ettringite only leads to expansive forces if it forms in small pores (~50 nm) (Taylor *et al.*, 2001).

Generally it can be said that the expansive formation of ettringite requires:

1 A source of alumina (e.g. calcium alumino monosulfate) distributed in small pores
2 A source of sulfate, which can locally produce conditions that are supersaturated with respect to ettringite formation.

In the formulated systems described above, such conditions may arise if there is an excess of calcium sulfate present, such that this does not all react to form ettringite during the initial reactions. Systems that are correctly formulated have been shown to be dimensionally stable over several years, even if subject to wetting and drying cycles.

2.11 Summary

The aim of this chapter has been to introduce the reader to the properties and applications of calcium aluminate cements. Such cements have now been produced for nearly a century and are the most important type of special cement in terms both of the quantity used and their range of applications. Due to their higher price, they should not be considered as a direct replacement for Portland cements. Today the major applications of these cements are in refractory concretes (described in a separate chapter) and as a component of pre-mixed binder systems used in the building industry, with lesser amounts being used in conventional concrete form.

In conventional concrete form it is important to take account of the phenomenon of conversion whereby the concrete attains its stable long-term strength. If the concrete is kept cool, this process may take several years, but when self-heating occurs during hardening, it is relatively rapid. It is important to maintain strict control on the water-to-cement ratio used to ensure adequate strength after conversion – it is recommended that the total w/c ratio should be at or below 0.4.

The principal properties of interest of CAC concrete are good resistance to chemical and mechanical attack and rapid strength development.

As a component in mixed binders, the presence of CAC contributes to the formation of ettringite, which can be controlled to give systems with rapid hardening, shrinkage control and rapid humidity reduction ('drying').

References

Amathieu, L., Bier, T.A. and Scrivener, K.L. (2001) Mechanisms of set acceleration of Portland cement. In Mangabhai, R.J. and Glasser, F.P. (eds), *Calcium Aluminate Cements 2001*. IOM Communications, London, pp. 303–317.

Code of Practice 114 (1957) and annex (1965) *The Structural Use of Reinforced Concrete in Buildings*. British Standards Institution, London.

Concrete Society (1997) *Calcium Aluminate Cement in Construction: A Re-assessment*. Technical Report No. 46, Concrete Society, Slough.

Cottin, B. and George, C.M. (1982) Reactivity of industrial aluminous cements: an analysis of the effect of curing conditions on strength development. *Proceedings of the International Seminar on Calcium Aluminates,* Turin 1982, pp. 167–70.

Damidot, D. (1998) Private communication.

Defargues, D. and Newton, T. (2000) The Ramsgate road tunnel. *Concrete* April, 43–45.

Deloye, F.-X., Lorang, B., Montgomery, R., Modercin, I. and Reymond, A. (1996) Comportement à long terme d'une liaison Portland-Fondu. *Bulletin des Laboratoires des Ponts et Chaussées–***202** (March/April), 51–59.

Ding, J., Fu, Y. and Beaudoin, J.J. (1996) Effect of different organic salts on conversion prevention in high-alumina cement products. *J. Adv. Cement Based Materials* **4**, 43–47.

French Standard (1991) *Use of Fondu aluminous cements in concrete structures* (annex). AFNOR P15–316.

Fryda, H., Scrivener, K.L., Chanvillard, G. and Féron, C. (2001) Relevance of laboratory tests to field applications of calcium aluminate cement concretes. In Mangabhai, R.J. and Glasser, F.P. (eds), *Calcium Aluminate Cements 2001*. IOM Communications, London, pp. 227–246.

Fryda, H. and Scrivener, K.L. (2002) User friendly calcium aluminate cement concretes for specialist applications. In *Proceedings International Conference on the Performance of Construction Materials in the New Millennium*, Cairo, Eygpt, February.

Füllmann, T., Walenta, G., Bier, T., Espinosa, B. and Scrivener, K.L. (1999) Quantitative Reitveld analysis of calcium aluminate cement. *World Cement* **30**(6).

French, P.J., Montgomery, R.G.J. and Robson, T.D. (1971) High strength concrete within the hour. *Concrete* August, 3.

George, C.M. (1976) The structural use of high alumina cement concrete. *Revue des Materiaux et Construction* 201–209.

George, C.M. and Montgomery, R.G.J. (1992) Calcium aluminate cement concrete: durability and conversion. A fresh look at an old subject. *Materiales de Construccíon* **42**, (228), 33.

Goyns, A. (2001) Calcium aluminate linings for cost-effective sewers. In Mangabhai, R.J. and Glasser, F.P. (eds), *Calcium Aluminate Cements 2001*. IOM Communications, London, pp. 617–631.

Lamour, V.H.R., Monteiro, P.J.M., Scrivener, K.L. and Fryda, H. (2001) Mechanical properties of calcium aluminate cement concretes. In Mangabhai, R.J. and Glasser, F.P. (eds), *Calcium Aluminate Cements 2001*. IOM Communications, London, 199–213.

Letourneux, R. and Scrivener, K.L. (1999) The resistance of calcium aluminate cements to acid corrosion in wastwater applications. In Dhir, R.K. and Dyer, T.D. (eds), *Modern Concrete Materials: Binders additions and admixtures*. Thomas Telford, London, 275–283.

Mácias, A., Kindness, A., Glasser, F.P. (1996) Corrosion behaviour of steel in high alumina cement mortar cured at 5, 25 and 55°C: chemical and physical factors. *J. Mat. Sci.* **31**, 2279–2289.

Majumdar, A.J. and Singh, B. (1992) Properties of some blended high-alumina cements. *Cement and Concrete Research* **22**, 1101–1114.

Mangabhai, R.J. and Glasser, F.P. (eds) (2001) *Calcium Aluminate Cements 2001*. IOM Communications, London.

Osborne, G.J. (1994) A rapid hardening cement based on high-alumina cement. *Proc. Inst. Civ. Eng., Structures and Buildings* **104**, 93–100.

Saucier, F., Scrivener, K.L., Hélard, L. and Gaudry, L. (2001) Calcium aluminates cement based concretes for hydraulic structures: resistance to erosion, abrasion & cavitation. *Proc. 3rd Int. Conf. Concrete in Severe Conditions: Environment and Loading, CONSEC'01*, Vancouver, Canada, June, Vol II, 1562–1569.

Scrivener, K.L., Lewis, M. and Houghton, J. (1997a). Microstructural investigation of calcium aluminate cement concrete from structures. *Proc 10th ICCC*, Göteborg, paper 4iv027.

Scrivener, K.L., Rettel, A., Beal, L. and Bier, T. (1997b) Effect of CO_2 and humidity on the mechanical properties of a formulated product containing calcium aluminate cement. *Proc. IBAUSIL 1997*, I-0745–0752, University of Weimar.

Scrivener, K.L. and Capmas, A. (1998) Calcium aluminate cements. In Hewlett, P.J. (ed.), *F.M. Leas's Chemistry of Cement and Concrete*, 4th edn. Arnold, London, 709–778.

Scrivener, K.L., Cabiron, J-L. and Letourneaux, R. (1999) High performance concrete based on calcium aluminate cements. *Cem. Concr. Res.* **29**, 1215–1223.

Taylor, H.F.W., Famy, C. and Scrivener, K.L. (2001). Delayed ettringite formation, *Cement and Concrete Research* **31**(5), 683–693.

Touzo, B., Glotter, A. and Scrivener, K.L. (2001) Mineralogical composition of fondu revisited. In Mangabhai, R.J. and Glasser, F.P. (eds) *Calcium Aluminate Cements 2001*. IOM Communications, London, 129–138.

Zhou, Q. and Glasser, F.P. (2001) *Thermal stability and decompositions mechanisms of ettringite at <120°C. Cement and Concrete Research* **31**, 1333–1339.

PART 2

Cementitious additions

3

Cementitious additions

Robert Lewis, Lindon Sear,
Peter Wainwright and Ray Ryle

This chapter covers the production and use in concrete of the principal cementitious additions, namely, pulverized fuel ash and natural pozzolanas, ground granulated blastfurnace slag, silica fume, metakaolin and limestone. It is recommended that this chapter be read in conjunction with Chapter 1.

3.1 The pozzolanic reaction and concrete

A pozzolana is a natural or artificial material containing silica in a reactive form. By themselves, pozzolanas have little or no cementitious value. However, in a finely divided form and in the presence of moisture they will chemically react with alkalis to form cementing compounds. Pozzolanas must be finely divided in order to expose a large surface area to the alkali solutions for the reaction to proceed. Examples of pozzolanic materials are volcanic ash, pumice, opaline shales, burnt clay and fly ash. The silica in a pozzolana has to be amorphous, or glassy, to be reactive. Fly ash from a coal-fired power station is a pozzolana that results in low-permeability concrete, which is more durable and able to resist the ingress of deleterious chemicals.

3.2 Fly ash as a cementitious addition to concrete

The first reference to the idea of utilizing coal fly ash in concrete was by McMillan and Powers in 1934 and in subsequent research (Davis *et al.*, 1935, 1937). In the late 1940s,

UK research was carried out (Fulton and Marshall, 1956) which led to the construction of the Lednock, Clatworthy and Lubreoch Dams during the 1950s with fly ash as a partial cementitious material. These structures are still in excellent condition, after some 50 years.

When coal burns in a power station furnace between 1250°C and 1600°C, the incombustible materials coalesce to form spherical glassy droplets of silica (SiO_2), alumina (Al_2O_3), iron oxide (Fe_2O_3) and other minor constituents. When fly ash is added to concrete the pozzolanic reaction occurs between the silica glass (SiO_2) and the calcium hydroxide ($Ca(OH)_2$) or lime, this is a by-product of the hydration of Portland cement. The hydration products produced fill the interstitial pores reducing the permeability of the matrix. Roy (1987) states 'the reaction products are highly complex involving phase solubility, synergetic accelerating and retarding effects of multiphase, multi-particle materials and the surface effects at the solid liquid interface'. The reaction products formed differ from the products found in Portland cement-only concretes. A very much finer pore structure is produced with time presuming there is access to water to maintain the hydration process. Dhir *et al.* (1986) have also demonstrated that the addition of fly ash improves the dispersion of the Portland cement particles, improving their reactivity.

Figures 3.1 to 3.3 show that $Ca(OH)_2$ (hydrated lime) is produced by the reaction of the Portland cement and water. Due to its limited solubility, particles of hydrated lime form within interstitial spaces. With a continuing supply of moisture, the lime reacts with the fly ash pozzolanically, producing additional hydration products of a fine pore structure. The pozzolanic reaction is as in

Calcium hydroxide + silica = tricalcium silicate + water

$$3Ca(OH)_2 + SiO_2 = 3CaO \cdot SiO_2 + 3H_2O$$

Figure 3.4, from Cabrera and Plowman (1981), shows the depletion of calcium hydroxide (lime) with time and how this reaction affects the long-term gain in strength of fly ash concrete (A) compared to a PC concrete control (B). The reaction takes place both within the pores of the cement paste and on the surface of fly ash particles. Despite the pozolanic

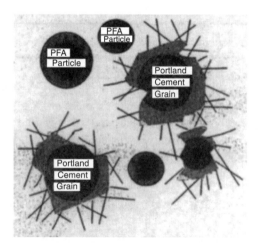

Figure 3.1 Hydration products of Portland cement.

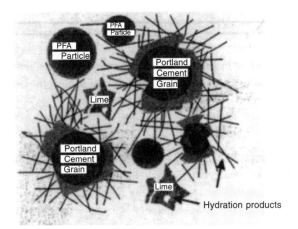

Figure 3.2 Lime is formed as a by-product of hydration.

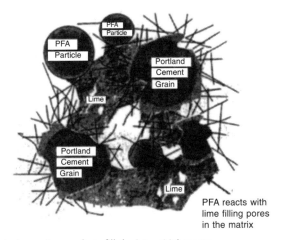

Figure 3.3 The pozzolanic reaction products fill the interstitial spaces.

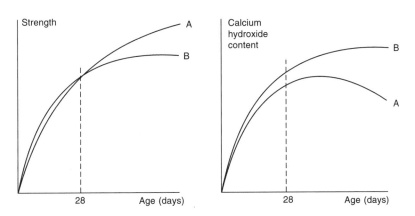

Figure 3.4 Influence of calcium hydroxide on strength development (Cabrera and Flowman, 1981).

reaction reducing the available hydrated lime in the pore solution, there is sufficient remaining to maintain a high pH.

Little pozzolanic reaction occurs during the first 24 hours at 20°C. Thus for a given cementitious content, with increasing fly ash content, lower early strengths are achieved. Taylor (1997) explains the hydration processes involved in some detail. The presence of fly ash retards the reaction of alite within the Portland cement at early ages. However, alite production is accelerated in the middle stages due to the provision of nucleation sites on the surface of the fly ash particles. The calcium hydroxide etches the surface of the glassy particles reacting with the SiO_2 or the Al_2O_3–SiO_2 framework. The hydration products formed reflect the composition of the fly ash with a low Ca/Si ratio. Clearly the surface exposed for reaction is greater, the finer the fly ash; additionally, the higher the temperature, the greater the reaction rate. At later ages the contribution of fly ash to strength gain increases greatly, provided there is adequate moisture to continue the reaction process.

3.2.1 The particle shape and density

Fly ash particles less than 50 μm are generally spherical (Figure 3.5) and the larger sizes tend to be more irregular. The spherical particles confer significant benefits to the fluidity of the concrete in a plastic state by optimizing the packing of particles. The fly ash spheres appear to act as 'ball bearings' within the concrete reducing the amount of water required for a given workability. In general the finer a fly ash the greater the water-reducing effect as in Figure 3.6 (Owens, 1979). Visually fly ash concrete may appear to be very cohesive, until some form of compactive effort is applied. Any reduction in water content reduces bleeding and drying shrinkage.

Figure 3.5 Fly ash particles are spherical.

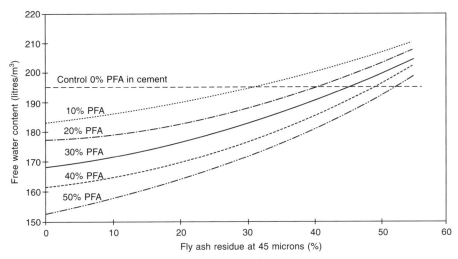

Figure 3.6 Finer fly ash and/or more fly ash reduces water content (Owens, 1979).

When relatively coarse fly ash, i.e. 45 μm residue >12 per cent, is interground with clinker or ground separately, the water requirement of concrete is markedly reduced according to Monk (1983). The grinding action appears to break down agglomerates and porous particles, but has little influence on fine glassy spherical particles, smaller than approximately 20 μm (Paya *et al.*, 1996).

The particle density of fly ash is typically 2300 kg/m³, significantly lower than for Portland cement at 3120 kg/m³. Therefore, for a given mass of Portland cement a direct mass substitution of fly ash gives a greater volume of cementitious material. The mix design for the concrete should be adjusted in comparison to a Portland cement of the same binder content to allow for the increased volume of fine material.

For a given 28-day strength the higher cementitious content needed in comparison with Portland cement concrete and lower water content required for fly ash-based concretes can give significantly higher quality surface finishes.

3.2.2 Variability of fly ash

Controlled fineness fly ash has been available in the UK since 1975 and over recent years, the majority of fly ash/PFA supplied for concrete complies with BS 3829 Part 1. This British standard limits the percentage retained on the 45 μm sieve to 12 per cent. The typical range of fineness found is < ± 5 per cent. This standard by default restricts the variability of the PFA. In practice in order to achieve the required fineness, PFA to BS 3892 Part 1 is classified using air-swept cyclones. The typical range of fineness found is ± 3 per cent. BS EN 450 permits a wider range of variation of ± 10 per cent on the declared mean fineness range, which is up to a maximum of 40 per cent retained on the 45 μm sieve.

The variability of the fineness may adversely influence the variability of the concrete, reducing the commercial viability of using fly ash. However, Dhir *et al.* (1981) concluded that LOI and fineness are only 'useful indicators' of fly ash performance. Dhir shows that the differing characteristics of seven cements produce a greater influence on concrete performance than the characteristics of eight differing fly ashes. He suggests that the fineness limits set in BS 3892 Part 1 are too low and that a limit of 20–25 per cent is more realistic. Matthews and Gutt (1978) report on the effects of fineness, water reduction, etc. on a range of ashes with up to 18 per cent retained on the 45 μm sieve. At a fixed water/cement ratio, the range of strengths found indicates coefficients of variation of 17 per cent at 365 days and 9 per cent at 28 days.

3.2.3 Fineness, water demand and pozzolanic activity

In one sense, fineness is important simply because a smaller particle size means a greater surface area will be exposed to the alkaline environment within the concrete. With a single source of fly ash, the strength performance can be related to water demand and fineness as in Figure 3.7, from Owens (1979).

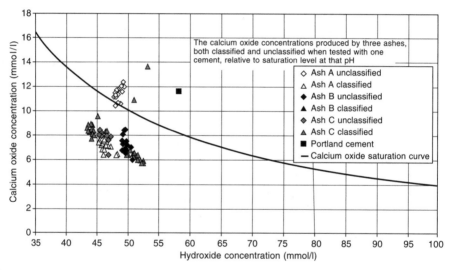

Figure 3.7 Hydroxide concentration reduction due to pozzolanicity of various fly ashes.

BS EN 196-5 details a method of determining the pozzolanicity of a fly ash by comparing the hydroxide concentration of a mortar containing fly ash against the theoretical hydroxide content saturation possible. The results of a testing programme carried out 1999 (NUSTONE Environmental Trust, 2000) clearly show the differences between sources (Figure 3.8) as well as the improved reactivity of classified fly ash.

The particle shape and finer fractions of fly ash are capable of reducing the water content needed for a given workability as in Figure 3.7. These effects are felt to be due to void filling on a microscopic scale replacing water within the concrete mix. Dewar's (1986) mix design system correlates with the water reductions found in practice when using fly ash. Where a water reduction is found this partially contributes to the relative strength performance of the cement/fly ash combination by acting as a solid particulate plasticizer.

The strength of fly ash concrete will depend on whether a water reduction is achieved, plus the pozzolanic performance of the cement/fly ash combination. While finer fly ashes

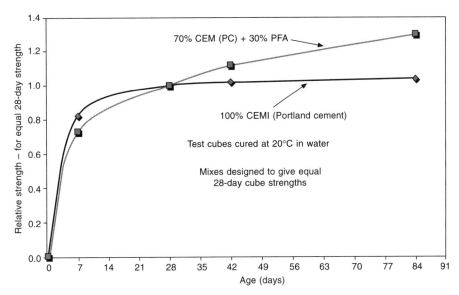

Figure 3.8 At 20°C concrete containing 30 per cent fly ash continues to gain significant strength.

are assumed to improve strength significantly with time, not all researchers find this true (Brown, 1980). Typically, strength development graphs of standard cubes with and without fly ash are shown in Figure 3.9. The graph also clearly illustrates what effect inclusion of fly ash in concrete has on the development of strength at early ages. Where higher curing temperatures are encountered, as in thick sections, significantly higher *in-situ* strength can be achieved than in test cubes cured at 20°C. Figure 3.10 shows 30 MPa grade, 1.5 m cubic concrete specimens (Concrete Society, 1998), that have achieved considerably higher *in-situ* strengths than the 28-day, 20°C cube strengths. These were insulated on five sides to recreate thick concrete sections.

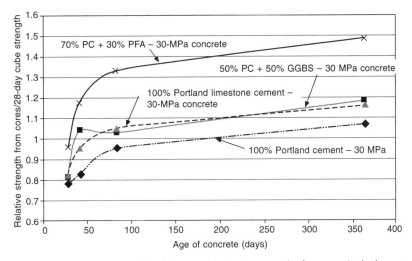

Figure 3.9 30 per cent PFA can significantly improve the *in-situ* strength of concrete in the longer term.

3.2.4 Heat of hydration

Development of concrete mix design has seen an increase in the proportion of cement being replaced by fly ash. Fly ash is able to reduce the heat of hydration very effectively, as in Figure 3.10 by Woolley and Conlin (1989).

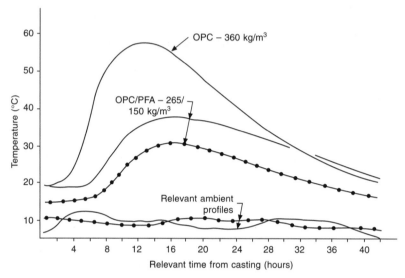

Figure 3.10 Heat of hydration with time (Woolley and Coulin, 1989).

The hydration of Portland cement compounds is exothermic, typically with 500 Joules per gram being liberated. The introduction of fly ash to replace a proportion of cement in concrete influences temperature rise during the hydration period. However, the rate of pozzolanic reaction increases with increasing temperature, however the peak temperatures in fly ash concrete are significantly lower than equivalent PC concretes. Figure 3.11 by Bamforth (1984) gives adiabatic temperature curves (°C per 100 kg of cement) for various cement contents.

3.2.5 Setting time and formwork striking times

Using fly ash in concrete will increase the setting time compared with an equivalent grade of PC concrete. There is a period before the hydration of fly ash concrete commences, but is has been shown by Woolley and Cabrera (1991) that the gain in strength, once hydration has started, is greater for fly ash concrete. When 30 per cent fly ash is used to replace PC in a mix, the setting time may be increased by up to 2 hours. This increased setting time reduces the rate of workability loss. However, it may result in finishing difficulties in periods of low temperature. In compensation, it will reduce the incidence of cold joints in the plastic concrete.

Formwork striking times at lower ambient temperatures may have to be extended in comparison to PC concrete, especially with thin sections. However, in practice vertical formwork striking times can be extended without this affecting site routines, e.g. the formwork is struck the following day. For soffit formwork, greater care has to be taken.

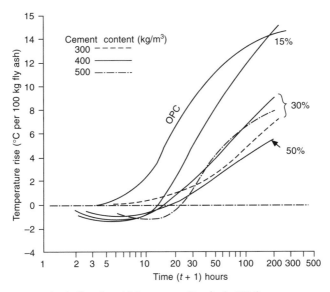

Figure 3.11 Temperature rise in fly ash and PC concretes (Bamforth, 1984).

Reference should be made to BS 8110 (BSI) for recommended striking times. Temperature-matched curing can be used to ensure that sufficient *in-situ* strength has been achieved while allowing for the concrete curing conditions.

3.2.6 Elastic modulus

The elastic modulus of fly ash concrete is generally equal to or slightly better than that for an equivalent grade of concrete. Figure 3.12 shows the elastic modulus and strength from Dhir *et al.* (1986) for concrete cured at different regimes.

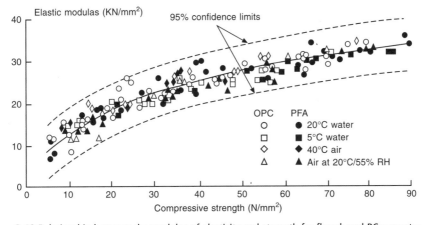

Figure 3.12 Relationship between the modulus of elasticity and strength for fly ash and PC concretes (Dhir *et al.*, 1986).

3.2.7 Creep

The greater long-term strength for fly ash concretes gives lower creep values, particularly under conditions of no moisture loss. These conditions may be found in concrete remote from the cover zone of a structure. Where significant drying is permitted the strength gain may be negligible and creep of OPC and fly ash concretes would be similar. Figure 3.13 shows the creep of fly ash and OPC concrete loaded to different stress levels ((Dhir, *et al.*, 1986).

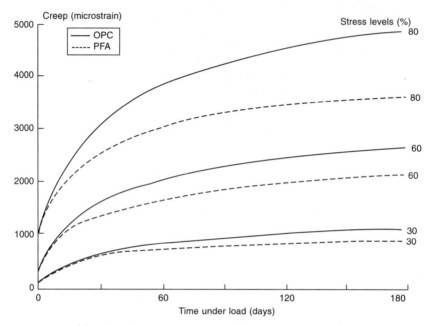

Figure 3.13 Creep of fly ash and PC concrete loaded to different stress levels (Dhir *et al.*, 1986).

3.2.8 Tensile strain capacity

Tensile strain capacity of fly ash concretes has been found to be marginally lower, and these concretes exhibit slightly more brittle characteristics (Browne, 1984). There is possibly a greater risk of early thermal cracking for given temperature drop, partially offsetting the benefits of lower heat of hydration in the fly ash concrete.

3.2.9 Coefficient of thermal expansion

The type of coarse aggregate used largely influences the coefficient of thermal expansion of concrete. Replacement of a proportion of cement with fly ash will have little effect on this property (Gifford and Ward, 1982).

3.2.10 Curing

Hydration reactions between cement and water provide the mechanism for the hardening of concrete. The degree of hydration dictates strength development and all aspects of durability. If concrete is allowed to dry out hydration will cease prematurely. Fly ash concrete has slower hydration rates and the lack of adequate curing will affect the final product. Thicker sections are less vulnerable than thin concrete sections because heat of hydration will promote the pozzolanic reaction.

3.2.11 Concrete durability and fly ash

Keck and Riggs (1997) quote Ed Abdun-Nur who said, 'concrete, which does not contain fly ash, belongs in a museum'. They list the benefits of using fly ash, e.g. low permeability, resistance to sulfate attack, alkali–silica reaction and heat of hydration giving references to their conclusions. Thomas (1990) reported that cores taken from 30-year-old structures in general have performed well; such structures must have been made with unclassified fly ashes over which minimal control of fineness, LOI, etc. would have been exercised. Berry and Malhotra review the performance of fly ash concretes in detail. They show the beneficial properties of using fly ash for sulfate resistance, chloride penetration, permeability, etc. Consideration of some specific durability aspects follows.

Penetration of concrete by fluids or gases may adversely affect its durability. The degree of penetration depends on the permeability of the concrete, and since permeability is a flow property it relates to the ease with which a fluid or gas passes through it under the action of a pressure differential. Porosity is a volume property, representing the content of pores irrespective of whether they are inter-connected and may/may not allow the passage of a fluid or gas.

3.2.12 Alkali–silica reaction

Alkali–silica reaction (ASR) (Concrete Society Technical Report 30, 1999) is potentially a very disruptive reaction within concrete. ASR involves the higher pH alkalis such as sodium and potassium hydroxides reacting with silica, usually within the aggregates, producing gel. This gel has a high capacity for absorbing water from the pore solution causing expansion and disruption of the concrete. Some greywacke aggregates have been found particularly susceptible to ASR. The main source of the alkalis is usually the Portland cement or external sources. Fly ash does contain some sodium and potassium alkalis but these are mainly held in the glassy structure and not readily available. Typically, only some 16–20 per cent of the total sodium and potassium alkalis in fly ash are water-soluble.

Many researchers have shown that fly ash is capable of preventing ASR. Oddly, the glass in fly ash is itself in a highly reactive fine form of silica. It has been found that the ratio of reactive alkalis to surface area of reactive silica is important in ASR. A pessimum ratio exists where the greatest expansion will occur. However, by adding more silica that is reactive plus the effect of the dilution of the alkalis mean that disruptive ASR cannot occur. The recommendations (BSI 5328 amendment 10365, 1999) within the UK require a minimum of 25 per cent BS 3892 Part 1 fly ash to prevent ASR. For coarser fly ashes, a minimum of 30 per cent fly ash may be required to ensure sufficient surface area to prevent ASR. Small quantities of fine fly ash with low-reactivity aggregates and sufficient alkalis may be more susceptible to ASR if the pessimum silica–alkali ratio is approached. Even when total alkalis within the concrete are as high as 5 kg/m^3, fly ash has been found (Alasali and Malhotra, 1991) able to prevent ASR. The addition of fly ash reduces the pH of the pore solution to below 13 at which point ASR cannot occur. The use of low-alkali cements has a similar effect. However, the detailed mechanisms by which fly ash prevents ASR are complex and imperfectly understood.

The ACI *Manual of Concrete Practice* (1994) suggests few restrictions on the effectiveness of fly ashes. It states that 'The use of adequate amounts of some fly ashes can reduce the amount of aggregate reaction'. It is later suggested that ashes only have to comply with ASTM C618, which permits a wide range of fly ashes. Fournier and Malhotra (1997) investigated the ability of a range of fly ashes to prevent ASR. The activity index (AI) was found to affect the ASR performance. However, there was no correlation between fineness and AI. Nant-y-Moch, Dinas and Cwm Rheidol Dams are excellent examples of fly ash preventing ASR constructed about the same time using the same aggregates. The Nant-y-Moch and Cwm Rheidol dams used 'run of station' fly ash as part of the cement. These dams have performed well, in comparison to the Portland cement-only Dinas dam, which has some evidence of ASR cracking.

3.2.13 Carbonation of concrete

As fly ash pozzolanically reacts with lime, this potentially reduces the lime available to maintain the pH within the pore solution. However, fly ash does reduce the permeability of the concrete dramatically when the concrete is properly designed and cured. When designing concretes for equal 28-day strength the slow reaction rate of fly ash usually means that the total cementitious material is often increased. This increase partially compensates for the reduction in available lime. Coupling this with the lower permeability leads to the result that the carbonation of fly ash concrete is not significantly different from Portland cement-only concrete of the same grade (28 days) (Concrete Society, 1991). The ACI *Manual of Concrete Practice* (1994) confirms this view: 'Despite the concerns that the pozzolanic action of fly ash reduces the pH of concrete researchers have found that an alkaline environment very similar to that in concrete without fly ash remains to preserve the passivity of the steel'.

Carbonation is a complex function of permeability and available lime. With properly designed, cured and compacted fly ash concrete, carbonation is not significantly different from other types of concrete. With extended curing and the low heat of hydration properties of fly ash concrete, the resulting low permeability may more than compensate for the reduced lime contents.

3.2.14 Sea water and chloride attack on reinforcing

Chlorides can come from many sources. First, it is important to restrict the chloride content in the constituents of the concrete to a minimum and guidance is given within various standards, e.g. BS 5328, BS 8110 and BS 8500. Chlorides from external sources are normally from seawater and de-icing salts used on roads. Tri-calcium aluminate (C_3A), a compound found in Portland cement, is able to bind chloride ions forming calcium chloro-aluminate. Similarly, tetra calcium alumino ferrite (C_4AF) can also reduce the mobility of chloride ions forming calcium chloro ferrite. Fly ash also contains oxides of alumina, which are able to bind chloride ions.

One highly effective way of reducing chloride ingress is to lower the permeability of the concrete. When fly ash is added to concrete, the permeability is significantly lower as in Figure 3.14 from Owens (1979). This reduces the transportation rate of chloride ions, and many other materials, into the concrete.

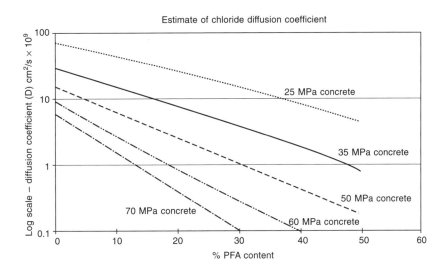

Figure 3.14 By increasing the strength and fly ash content of concrete, the permeability reduces significantly (Owens, 1979).

3.2.15 Freeze–thaw damage

Fly ash concrete of the same strength has a similar resistance to freeze–thaw attack as Portland cement concrete. Dhir *et al.* (1987) reported on the freeze–thaw properties of concrete containing fly ash. They used nine fly ashes of various sieve residues. All mixes were designed to give equal 28-day strengths. Freeze–thaw was assessed using an 8-hour cycle of 4 hours at –20°C and 4 hours at +5°C. Ultrasonic pulse velocity (UPV) and changes in length were used for the assessment of freeze–thaw performance. They found, as have other researchers, that adding fly ash reduces freeze–thaw resistance unless air entrainment is used. However, Dhir found that only 1.0 per cent of air entrained in the concrete gave superior performance compared with plain concrete irrespective of the cement type.

It is clear that the less permeable and denser the mortar matrix is within the concrete, the less space is available to relieve the pressures associated with the expansion of freezing water. The additional hydration products created by the pozzolanic reaction result in fly ash being less frost resistant unless pressure-releasing voids are artificially created by the addition of air entrainment. In all low-permeability concretes, irrespective of the fineness or chemistry of the fly ash or other constituents, freeze–thaw resistance is given by air entrainment. The air bubble structure is able to relieve expansive stresses of the freezing water presuming proper curing regimes have been adhered to.

3.2.16 Sulfate attack

Fly ash concrete can increase the resistance to sulfate attack compared with a CEM I concrete of similar grade. Deterioration due to sulfate penetration results from the expansive pressures originated by the formation of secondary gypsum and ettringnite. The beneficial effects of fly ash have been attributed to a reduction of pore size slowing penetration of sulfate ions. Less calcium hydroxide is also availble for the formation of gypsum.

The smaller pore size of fly ash concrete reduces the volume of ettringnite that may be formed. One of the major constituents of cement that is prone to sulfate attack, tricalcium aluminate (C_3A), is diluted since a proportion of it will have reacted with the sulfates within the fly ash at an early age. Building Research Establishment Special Digest 1 (BRE, 1991) discusses the factors responsible for sulfate and acid attack on concrete below ground and recommends the type of cement and quality of curing to provide resistance. Concrete made with combinations of Portland cement and BS 3892: Part 1 PFA, where the fly ash content lies between 25 per cent and 40 per cent has good sulfate-resisting properties.

Seawater contains sulfates and attacks concrete through chemical action. Crystallization of salts in pores of the concrete may also result in disruption. This is a particular problem between tide marks subject to alternate wetting and drying. Presence of a large quantity of chlorides in seawater inhibits the expansion experienced where groundwater sulfates have constituted the attack.

3.2.17 Thaumasite form of sulfate attack

There has much been said about thaumasite in the UK since the discovery of a number of damaged structures on a UK motorway. Bensted (1988) describes the chemistry of the thaumasite reaction. Thaumasite attack occurs at temperatures below 15°C and requires the presence of calcium carbonate. Limestone aggregates are a source of calcium carbonate, with oolitic limestone appearing to be the most reactive form. Sulfates react with the calcium carbonate and the C_3S and C_2S hydrates forming thaumasite. As these are the strength-giving phases of the cement, their removal results in the concrete disintegrating to a white powdery sludge-like material. This reaction is not expansive and may not be easily detected below the ground. At the time of writing, a number of research projects remain outstanding, which should answer some of the questions this reaction poses.

Burton (1980) carried a series of mixes to determine the relative performance of concretes made with Portland cement, sulfate resisting Portland cement (SRPC) and Portland/fly ash blended cement concretes when exposed to sulfate attack. The sulfate solutions were not heated and the tanks were in an unheated, external storeroom. Interestingly the Portland cement mixes all showed signs of deterioration after six months in magnesium sulfate solution but only slight deterioration in sodium sulfate solution. However, after 5 years immersion total disintegration of all the PC samples occurred. The oolitic limestone mixes suffered between 22.2 and 42.5 per cent weight loss after one year in magnesium sulfate and the crushed carboniferous limestone mixes 14.3 to 27.3 per cent weight loss. This indicates the effect of aggregate type on the rate of deterioration. Though thaumasite was not well understood at the time of the project, a re-evaluation of the data and photographs suggests the deterioration found with both the PC and SRPC mixes was probably due to the thaumasite form of sulfate attack.

Bensted believes the addition of fly ash should reduce the problems with thaumasite simply because the permeability of fly ash concrete should prevent sulfate in solution from penetrating to any depth.

3.2.18 Resistance to acids

All cements containing lime are susceptible to attack by acids. In acidic solutions, where the pH is less than 3.5, erosion of the cement matrix will occur. Moorland waters with low hardness, containing dissolved carbon dioxide and with pH values in the range 4–7 may be aggressive to concrete. The pure water of melting ice and condensation contain carbon dioxide and will dissolve calcium hydroxide in cement causing erosion. In these situations, the quality of concrete assumes a greater importance.

3.3 Fly ash in special concretes

3.3.1 Roller-compacted concrete

High fly ash content roller-compacted concrete depends on achieving the optimum packing of all constituents in the concrete (Dunstan, 1981). That means that all voids should be filled. Optimizing the coarse aggregate content of a mix is common, but filling of the voids in the mortar fraction is rarely considered. For mix design, the paste fraction is the absolute volume of the cementitious materials and free water. The mortar fraction is the absolute volume of the fine aggregate and the paste fraction. Adopting these mix design principles has resulted in roller-compacted concretes with 60–80 per cent of fly ash being used in such applications.

3.3.2 High fly ash content concrete

Mix design principles developed for roller-compacted concrete have been used for concrete containing high volumes of fly ash. These may be compacted by immersion vibration plant (Cabrera and Atis, 1998). Adopting the principle of minimum voids in the paste, mortar and aggregate, mix designs have been successfully used for concrete placed to floor slabs, structural basements and walls. These concretes have 40–60 per cant of the cementitious volume as fly ash. Other work, using the maximum packing, minimum porosity principle (Cabrera *et al.*, 1984) have similarly been designed for structural concrete with fly ash making up to 70 per cent of the cementitious content by weight.

3.3.3 Sprayed concrete with pfa

Limited experimental work has shown that a major reduction in rebound can be achieved by replacing a proportion of cement with fly ash, or by using a blended fly ash–cement component. Using the 'Dry' process of spraying and replacing 30 per cent of the cement content with fly ash, 30 per cent reduction of generated rebound was recorded. At the same time, the hardened concrete revealed a smaller pore size with no loss of compressive strength (Cabrera and Woolley, 1996). This finding has major potential for the replacement of a proportion of cement with pfa in sprayed concrete mixes.

3.4 Natural pozzolanas

Natural pozzolanas are glassy, amorphous materials that are normally classified into four groups:

1 Volcanic materials – these are rich in glass in an unaltered or partially altered state. They are found throughout the world in volcanic regions. They result from the explosion of magma that rapidly cooled by being quenched to form microporous glass. Often these materials consist of >50 per cent silica, followed by alumina, iron oxides and lime. They often have a high alkali content of up to 10 per cent.
2 Tuffs – where a volcanic glass has been partially or fully transformed into zeolitic compounds due to weathering. The range of the chemical composition is similar to volcanic materials. The loss on ignition is an indication of the degree of transformation that has occurred due to weathering.
3 Sedimentary – these are rich in opaline diatoms. The diatoms are the skeletons of organisms that are mainly composed of opal. They have high silica content; however, these materials are often polluted with clay.
4 Diagenetic – These are rich in amorphous silica and derived from the weathering of siliceous rocks. They are normally high in silica and low in other oxides, depending on the mineral composition of the parent rocks.

With all pozzolanic materials, the gradation and particle shape will have a significant effect on the water requirement of the concrete. Naturally occurring pozzolanas often have irregular shape that will increase water demand unless plasticizers are used to compensate.

3.5 The use of ggbs in concrete

3.5.1 Introduction

The hydraulic potential of ground granulated blastfurnace slag (ggbs) was first discovered as long ago as 1862 in Germany by Emil Langen. In 1865 commercial production of lime-activated ggbs began in Germany and around 1880 ggbs was first used in combination with Portland cement (PC). Since then it has been used extensively in many European countries, such as Holland, France and Germany. In the UK the first British Standard for Portland Blastfurnace Cement (PBFC) was produced in 1923.

3.5.2 Production of ggbs

Blastfurnace slag is produced as a by-product during the manufacture of iron in a blastfurnace. It results from the fusion of a limestone flux with ash from coke and the siliceous and aluminous residue remaining after the reduction and separation of the iron from the ore. To process the slag into a form suitable for use as a cementitious material one of two techniques can be employed, namely granulation or pelletization; with either technique it is essential that the slag be rapidly cooled to form a glassy disordered structure. If the slag is allowed to cool too slowly this allows a crystalline well-ordered structure to form which is stable and non-reactive. With granulation the stream of molten slag is forced over a weir into high-pressure water jets. This causes the slag to cool rapidly into glassy granules no larger than about 5 mm in diameter. In most modern granulators the water temperature is kept below about 50°C and the water slag ratio is normally between ten and twenty to one. Following granulation the granulate is dried and ground to cement fineness in a conventional cement clinker grinding mill.

Pelletization involves pouring the molten slag onto a water-cooled steel-rotating drum of approximately 1 m diameter. The drum has fins projecting from it, which throw the slag through the air inside a building where water is sprayed onto it thus causing it to cool rapidly. Pelletizing produces material from about 100 mm down to dust; the larger particles tend to be crystalline in nature and have little or no cementitious value. Particles larger than about 6 mm are therefore screened off and used as a lightweight aggregate in concrete and only the finer fraction (< 6 mm) used for the manufacture of ggbs. Granulation is much more efficient at producing material with a high glass content, but the capital costs of a granulator are about six times greater than that of a pelletizer. Both materials can be used as raw feed for ggbs and most Standards do not differentiate between them.

3.5.3 Chemical composition of ggbs

The chemical composition of slag will vary depending on the source of the raw materials and the blastfurnace conditions. The major oxides exist within the slag glass (formed as the result of rapid cooling) in the form of a network of calcium, silicon, aluminium and magnesium ions in disordered combination with oxygen. The oxide composition of ggbs is shown in Table 3.1 where it is compared with that of Portland cement.

Table 3.1 Typical oxide composition of UK Portland cement and ggbs

| Oxide | Composition (%) | |
	Portland cement	ggbs
C_aO	64	40
SiO_2	21	36
Al_2O_3	6.0	10
Fe_2O_3	3.0	0.5
MgO	1.5	8.0
SO_3	2.0	0.2
K_2O	0.8	0.7
Na_2O	0.5	0.4

3.5.4 Chemical reactions

In the presence of water ggbs will react very slowly but this reaction is so slow that on its own slag is of little practical use. Essentially the hydraulicity of the slag is locked within its glassy structure (i.e. it possesses latent hydraulicity) and in order to release this reactivity some form of 'activation' is required. The activators, which are commonly sulphates and/or alkalis, react chemically with the ggbs, and increase the pH of the system. Once a critical pH has been reached the glassy structure of the slag is 'disturbed', the reactivity is released and the slag will begin to react with the water producing its own cementitious gels. In practice activation is achieved by blending the ggbs with Portland cement as the latter contains both alkalis ($Ca(OH)_2$, NaOH and KOH) and sulphates. The chemical reactions that occur between the Portland cement, the water and the ggbs are very complex and are summarized below.

Hydration mechanism

- Primary reactions

$$OPC + water \rightarrow C.S.H. + Ca(OH)_2 \ \left.\begin{matrix} \\ NaOH \\ KOH \end{matrix}\right\} \qquad (1)$$

$$GGBS + water + Ca(OH)_2 \rightarrow C-(N,K)-S-H \qquad (2)$$
$$NaOH$$
$$KOH$$

- Secondary reactions:

   (i) OPC primary reaction products + ggbs
   (ii) OPC primary reaction products + ggbs primary reaction products.

Reaction rate

The rate of the chemical reactions is influenced by several factors including:

- The quantity and relative proportions of the two components: the lower the proportion of cement, the slower the reaction.
- Temperature: just as with Portland cement the rate of the reaction increases with an increase in temperature and vice versa
- The properties of the two components, in particular chemical composition and fineness.

Material properties

The most important properties of the slag are summarized in Tables 3.2 and 3.3 together with the limiting values taken from the appropriate standard.

Chemical composition Various hydraulic indices have been proposed to relate composition of the slag to hydraulicity; most of these imply an increase in hydraulicity with increasing CaO, MgO and a decrease with increasing SiO_2. BS 6699 uses the term chemical modulus or reactivity index (see Table 3.2) and states that $(CaO + MgO)/SiO_2$ should be greater than 1.0. In addition, as the CaO/SiO_2 ratio increases, the rate of reactivity of the slag also increases up to a limiting point when increasing CaO content makes granulation to a glass difficult, BS 6699 limits this ratio to a maximum vlaue of 1.4

Table 3.2 Limits for ggbs according to BS 6699

Property	Limit
Chemical modulus/Reactivity index $\dfrac{CaO + MgO}{SiO_2}$	≥ 1.0
Lime–silica ratio $\dfrac{CaO}{SiO_2}$	≤ 1.4
Glass count XRD	$\geq 67\%$
Fineness	$> 275 \ m^2/kg$
Moisture content	$\leq 1.0\%$

Table 3.3 Other properties of typical UK Portland cements and ggbs

Property	Portland cement	ggbs
Shape	Angular	Angular
RD	3.15	2.90
Colour	Grey	White

Fineness   Slag is normally ground finer than Portland cement to a fineness of $> 400 \ m^2/kg$; the finer the material, the more reactive it becomes.

Glass count

It is the non-crystalline (glassy) component of the slag that is potentially reactive; therefore the greater the proportion of glass, the more reactive the slag. The most common method of measuring glass count is by X-ray diffraction (XRD) and when using this method the glass count should be ≥ 67 per cent.

A summary of some of the other physical properties of slag is shown in Table 3.3 where comparisons are made with Portland cement. Because, like cement clinker, the slag granulate is ground its particle shape is angular although the surface texture tends to be somewhat smoother than cement. The relative density of the slag is slightly lower than that of cement (2.9 cf 3.1).

3.5.5 Blended cement production

The blended cement produced from the combination of Portland cement with ggbs can be manufactured by the intergrinding of the two components (i.e. clinker and granulated slag) in the ball mills or by blending the two components (i.e. PC and ggbs) as separate powers at the factory or by blending in the mixer. Intergrinding can lead to problems because the slag granulate is much harder than the clinker leading to preferential grinding of the cement clinker. Most cement producers in Europe use separate grinding followed by factory blending of the two powders. In the UK at present the favoured method of production is that of separate grinding combined with mixer blending.

The situation regarding Standards is a little confusing. There is a British Standard (BS 6699) covering the powder (i.e. ggbs) but as yet no equivalent European Standard and there is a European Standard (ENV 197-1 CEM II and CEM III) covering the slag cement which has replaced BS 146 and BS 4246.

Details on the limits of the two components (slag and clinker) for slag cements as given in ENV 197-1 are shown in Table 3.4. It can be seen from this table that it is possible to replace up to 95 per cent of the cement with ggbs (CEMIII Type C) although replacement levels of between 50–70 per cent are more typical for structural concrete applications.

Table 3.4 European Standard ENV 197-1 for slag cements

Components (%)	CEM II Portland slag cement		CEM III Blastfurnace cement		
	Type A	Type B	Type A	Type B	Type C
PC clinker	80–94	65–79	35–64	20–34	5–19
ggbs	6–20	21–35	36–65	66–80	81–95
Minor constituents	0–5	0–5	0–5	0–5	0–5

3.5.6 Properties of concrete made with slag cements

Plastic concrete
Water demand/workability   For concrete made with equal slump, a lower water content is required compared to Portland cement although the reductions are small and are no more than about 3 per cent. This reduction is related largely to the smoother surface texture of the slag particles and to the delay in the chemical reaction.

Stiffening times   Because the ggbs is slower to react with water than Portland cement its use is likely to lead to an increase in the stiffening times of the concrete. The extension in stiffening time will be greater at high replacement levels (above about 50 per cent) and at lower temperatures (below about 10°C). Figure 3.15 (Concrete Society, 1991) shows the influence of increasing additions of ggbs on the extension to setting times measured in accordance with BS 4550.

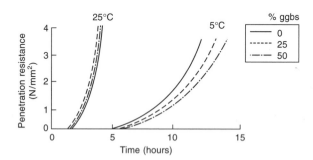

Figure 3.15 Influence of slag content and temperature on setting time (Concrete society, 1991).

Bleeding and settlement   The influence of slag on bleeding and settlement depends to some extent upon what basis comparisons are made. When testing at constant water/cement ratio the use of slag, in proportions greater than about 40 per cent, will invariably lead to increases in both bleeding and settlement. When, however, comparisons are made on the basis of equal 28-day strength the differences are not so marked as can be seen in Figure 3.16 (Wainwright, 1995). It should also be noted that there are marked differences between the bleed characteristics of Portland cements from different sources and in some cases these differences can be as great as those resulting from the addition of slag.

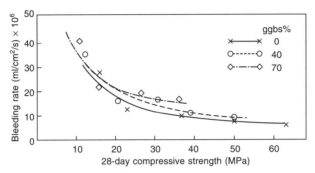

Figure 3.16 Influence of ggbs on bleeding at equal compressive strengths (Wainwright, 1995).

Heat of hydration and early age thermal cracking   The rate of heat evolution associated with ggbs is reduced as the proportion of slag is increased. This may be of benefit in large pours enabling greater heat dissipation and reduced temperature rise which will reduce the likelihood of thermal cracking. The actual temperature reductions that can be achieved in practice depend on many factors including: section size, cement content, proportion of slag, and chemical composition and fineness of the cementitious components. The benefits to thermal cracking from the reduced temperature rise may be offset to some extent by the lower creep characteristics of concretes containing ggbs (see below).

Figure 3.17 (Wainwright and Tolloczko, 1986) shows typical temperature–time profiles for concretes with and without slag and Figure 3.18 (Bamforth, 1984) shows the influence on temperature rise of different slag additions at different lift heights.

Hardened concrete
Compressive strength and strength development   Since ggbs hydrates more slowly than Portland cement the early rate of strength development of slag concretes is slower, the higher the slag content the slower the strength development. However, provided adequate moisture is available, the long-term strength of the slag concretes is likely to be higher. This higher later age strength is due in part to the prolonged hydration reaction of the slag cements and to the more dense hydrate structure that is formed as a result of the slower hydration reaction.

As with Portland cements the rate of the hydration reaction of slag cements is temperature dependent but slag cements have a higher activation energy than Portland cements and therefore their reaction rates are more sensitive to temperature change. This means that as the temperature increases the increase in the rate of gain of strength of slag cement is greater than that of Portland cement but the opposite is true as the temperature reduces. The influence of temperature on strength development is of significance when considering the behaviour of concrete *in situ*. In such situations the rate of strength development and

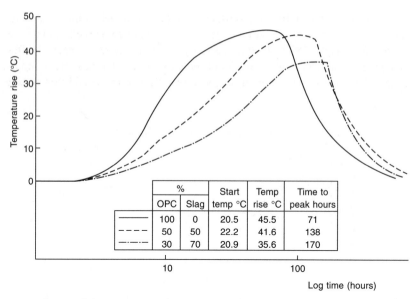

	%		Start	Temp	Time to
	OPC	Slag	temp °C	rise °C	peak hours
———	100	0	20.5	45.5	71
- - - -	50	50	22.2	41.6	138
-·-·-·	30	70	20.9	35.6	170

Figure 3.17 Influence of slag content on adiabatic temperature rise (Wainwright and Tolloczko, 1986).

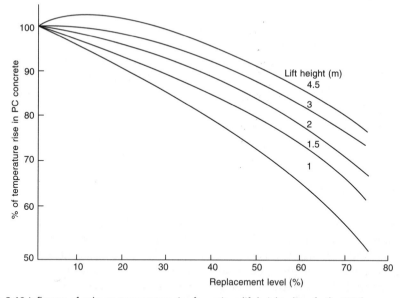

Figure 3.18 Influence of ggbs on temperature rise for various lift heights (Bamforth, 1984).

ultimate strength may be appreciably different from that indicted by standard cured cubes. In thick sections or cement-rich concrete the early temperature rise may well be in excess of 50°C as shown previously in Figure 3.3. The effect of this is to accelerate the early strength gain and to impair long-term development for concretes with and without ggbs. However, the amount by which the long-term strength gain is reduced is lower for slag cements than for Portland cements. Conversely in thin sections cast in cold weather conditions the slower rate of gain of strength may lead to some extension in formwork striking times.

The influence of temperature on strength development is shown in Figure 3.19 (Wainwright and Tolloczko, 1986) and Figure 3.20 (Swamy, 1986). Concretes which have been subjected to 'heat cycled' curing in Figure 3.19 have undergone a temperature cycle similar to that which they are likely to experience in a structure with relatively thick sections (i.e. similar, for example, to that shown in Figure 3.17). It is clear from Figure 3.19 that under heat cycled conditions the 70 per cent slag mix exceeded the strength of the PC mix after about 3 days yet under normal curing conditions it was still about 10 N/mm² below the PC mix after 6 months. In Figure 3.20 the specimens have been kept under constant temperature conditions, the mixes are not the same as those shown in Figure 3.19. Figure 3.20 illustrates how much more the strength of the slag mix is retarded at 5°C compared to the PC mix

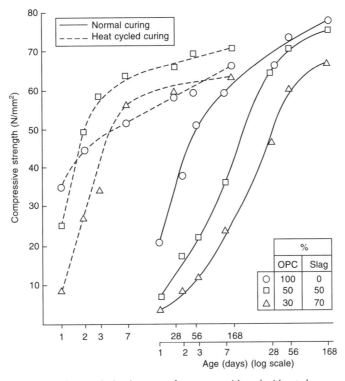

Figure 3.19 Comparison of strength development of concretes with and without slag, normal cured and heat cycled cured (Wainwright and Tolloczko, 1986).

Tensile strength   Compared to Portland cement slag cement concretes tend to have a slightly higher tensile strength for a given compressive strength although the differences are of little practical significance.

Elastic modulus   The use of ggbs will have the effect of increasing slightly the elastic modulus of the concrete for a given compressive strength when compared with Portland cement. However, as with tensile strength the magnitude of this difference is not large

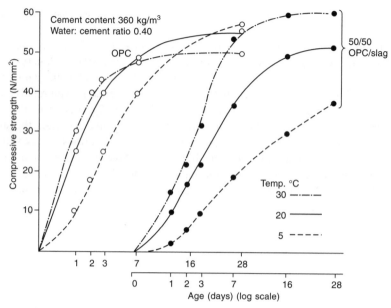

Figure 3.20 Influence of temperature on strength development for concretes with and without slag (Swamy, 1986).

and the relationship between strength and modulus for all both slag and Portland cements follows closely that given in BS 8110.

Creep Under conditions of no moisture loss the basic creep of concrete is reduced as the level of ggbs is increased. For high replacement levels (i.e. > 70 per cent ggbs) reductions as high as 50 per cent have been reported. The reductions in creep are generally associated with the greater strength gain at later ages of concretes containing slag. Under drying conditions this later age strength gain may be small and the differences in creep are likely to be less pronounced. For most practical situations where drying shrinkage is moderate the behaviour of concretes made from slag cements is likely to be similar to that of Portland cement concretes. Figures 3.21 and 3.22 (Neville and Brooks, 1975) show the influence of ggbs additions on both the basic and drying creep of concrete.

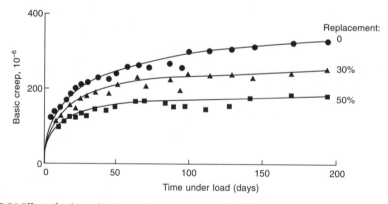

Figure 3.21 Effect of ggbs on basic creep in water at 22°C (Neville and Brooks, 1975).

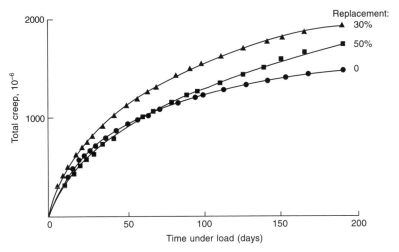

Figure 3.22 Effect of ggbs on total creep at 21°C and 60 per cent RH (Neville and Brooks, 1975).

Drying shrinkage   The use of ggbs has very little if any influence on the drying shrinkage of the concrete.

Other properties related to construction
Surface finish   The slight improvement in workability and small increase in paste volume of ggbs concretes generally makes it easier to achieve a good surface finish. In addition the colour of the concrete will be lighter than that of Portland cement concretes.

Formwork pressures   As a result of the extended setting times mentioned earlier the use of slag in concrete will lead to an increase in formwork pressures (Clear and Harrison, 1985). The increase in pressure will be particularly pronounced in high lifts (> 4 m) cast at low temperatures (< 5°C) and at low placing rates (< 0.5 m/h).

Formwork striking times   The slower rate of gain of early strength of concretes containing high levels of ggbs may require the extension of formwork striking times. In practice, however, the actual construction process often requires concrete to be cast one day and vertical formwork struck the next. In such cases it is quite likely that the minimum striking times will in any case be extended and that therefore the use of slag may not affect the actual construction process.

Curing   In situations where durability is recognized as a potential problem then it may be expected that concretes containing slag will require longer curing periods particularly in cold weather conditions and where thin sections are involved.

Durability
It is generally recognized that the durability of concrete is primarily related to its permeability/diffusion to liquids and gases and its resistance to penetration by ions such as sulfates and chlorides. Generally speaking, provided the concretes have been well-cured slag concretes are likely to be more durable than similar Portland cement concretes.

Permeability In well-cured concretes the inclusion of ggbs will be beneficial in terms of the long-term permeability and particularly so at higher temperatures. The reasons for this are:

1   The continued hydration which takes place well beyond 28 days in the slag concretes.
2   The overall finer pore structure associated with ggbs as shown in Figure 3.23 (Pigeon and Regourd, 1983), as the slag content increases, the number of smaller pores increases and the overall pore size distribution becomes finer even though there may be little change in porosity.
3   The influence of elevated temperature on pore structure is illustrated in Table 3.5 (Kumar *et al.*, 1987). In these tests conducted on cement pastes an increase in curing temperature resulted in a coarsening of the pore structure for Portland cements yet for the slag cements the pore structure was little affected.

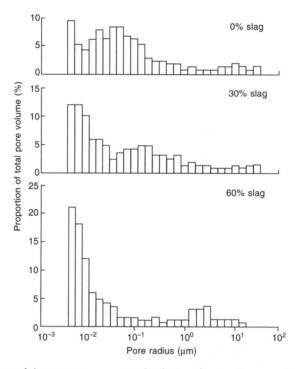

Figure 3.23 Influence of slag content on pore size distribution of pastes (Pigeon and Regourd, 1983).

Figure 3.24 (Grube, 1985) illustrates the problems associated with trying to relate small-scale laboratory tests data to full scale *in-situ* situations where the effects of temperature and the reserve of moisture may be significant. The results are from cast cylinders and from cores taken from walls 200 mm thick. In both cases the concrete was made to the same water/cement +ggbs ratio and storage conditions were either outdoors or in laboratory air at 20°C, 65 per cent RH. After only three days curing the cores made with the slag concrete exhibited lower permeabilities to oxygen than those made from OPC yet with the cast specimens even after 28 days the slag had only just reached parity with the PC.

Table 3.5 Influence of temperature and cement type on pore structure of cement pastes

Composition (by wt)	Age (days)	Porosity (%) Cured at			Medium pore size (nm) Cured at		
		27°C	38°C	60°C	27°C	38°C	60°C
Type I Portland	7	25.0	24.0	25.0	13.5	12.0	9.0
cement (OPC)	28	20.0	22.0	22.0	13.0	15.0	13.5
	90	17.0	19.0	24.5	10.5	12.0	17.5
65% GGBS	7	21.5	19.0	17.0	13.5	13.5	11.0
+	28	12.2	11.5	12.0	2.75	2.75	2.95
35% OPC	90	9.5	10.0	10.0	2.30	2.45	2.60
	180	8.0	8.0	8.0	2.35	2.40	2.50

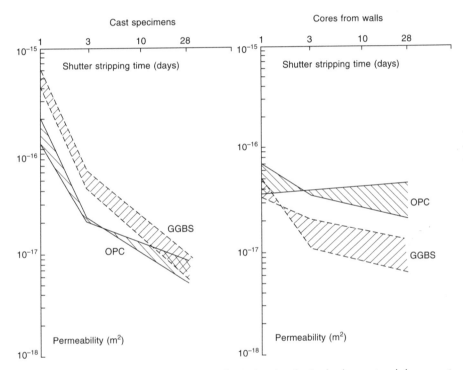

Figure 3.24 Relationship between permeability and stripping time for Portland cement and slag cement concretes (Grube, 1985).

Carbonation   The ability of the concrete to protect the steel from corrosion depends, among other factors, upon the extent to which the concrete in the cover zone has carbonated. The influence of additions of ggbs on carbonation has been the subject of much research and there still appears to be some disagreement as to its effects. The reasons for much of this debate appear to be related to the test procedures and conditions used in the studies and to the basis on which comparisons have been made.

Results obtained on cubes subjected to outdoor exposure conditions (Figure 3.25, Osborne 1986) indicate that, on the basis of equal 28-day compressive strength, there is a marginal increase in carbonation depths for ggbs concrete but for strengths above about

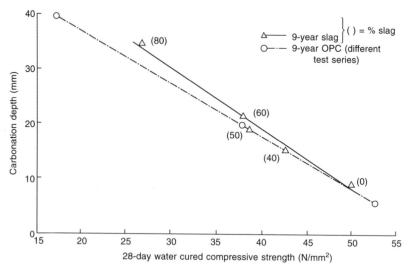

Figure 3.25 Influence of slag on relationship between carbonation depth and strength (Osborne, 1986).

30.0 N/mm^2 the differences are unlikely to be of practical significance. In other tests in which concretes were exposed at an early age (< 24 hours) the depth of carbonation was found to be largely controlled by the water/Portland cement ratio (Figure 3.26, Concrete Society, 1991). In such cases the depth of carbonation of the slag concretes was found to be significantly higher particularly at water/cement ratios above about 0.45. In contrast, extensive surveys of old concrete structures carried out on the Continent have shown no differences in carbonation depths for concretes with and without slag. Similar surveys carried out in the UK found an increase in carbonation depth at high replacement levels (> about 65 per cent) where the concrete was exposed to a sheltered environment but where the concrete was exposed to rainfall there was hardly any difference.

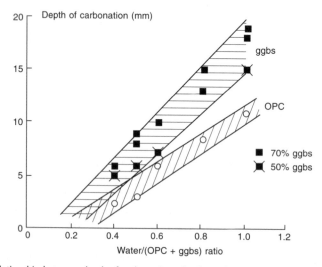

Figure 3.26 Relationship between depth of carbonation of ggbs and OPC concretes and water/(OPC + ggbs) ratio (Concrete Society, 1991).

Alkali–silica reaction (ASR)   The alkali–silica reaction is the most common form of alkali aggregate reaction and occurs as the result of a reaction between the alkaline pore fluid in the cement and siliceous minerals found in some aggregates. This reaction results in the formation of a calcium silicate gel which imbibes water producing a volume expansion and disruption of the concrete.

One effective way of reducing this expansion is by the use of ggbs as illustrated Figure 3.27 (Hobbs, 1982) which shows results of a laboratory study carried out in the UK. Although ggbs contains alkalis (in some cases quite a high level) their solubility is somewhat lower than those found in Portland cement. The mechanism by which these alkalis contribute to the alkalinity of the pore solution and therefore take part in the reaction is complex and is still not yet fully understood. In 1999 the Building Research Establishment in the UK published modified guidelines in the form of a Digest (BRE Digest 330, 1999) on ways to minimize the risk of attack from alkali–silica reaction. The Digest is a comprehensive document published in four parts and it is only intended here to highlight the main points contained in it as outlined below:

- Account must be taken of the reactivity of the aggregate, the aggregates being classified as: Low, Normal and High Reactivity.
- The use of a low alkali cement is recommended which is defined as a cement with an alkali content of ≤ 0.6 per cent Na_2O equivalent.

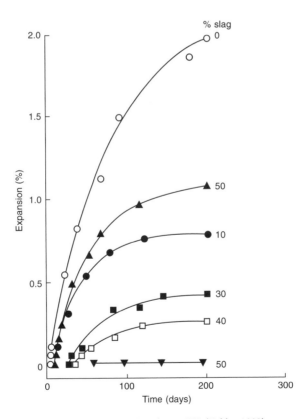

Figure 3.27 Influence of slag content on expansion due to ASR (Hobbs, 1982).

- One way of achieving a low alkali cement is by blending Portland cement with ggbs, the minimum recommended levels of slag being, 50 per cent or 40 per cent ggbs depending on aggregate reactivity.
- Where the minimum level of ggbs is used, no account need be taken of the alkali content of the slag when calculating the total alkali content of the concrete.
- The limit on total alkalis in the concrete ranges from 2.5–5.0 kg Na_2O equiv/m^3 depending on the reactivity of the aggregate.

The procedures for calculating the total alkali content in the concrete are far more complex than those given in previous guidelines. Readers are strongly advised to consult the Digest before attempting to carry out any calculations.

Sulfate resistance Concretes containing ggbs are acknowledged to have higher resistance to attack from sulfates than those made with only Portland cements. This improved resistance is related to the overall reduction in the C_3A content of the blended cement (ggbs contains no C_3A) and to the inherent reduction in permeability. It is generally accepted that provided the Al_2O_3 content of the slag is less than 15 per cent then cements containing at least 70 per cent ggbs can be considered comparable to sulfate-resisting Portland cement.

In the UK recommendations relating to resistance to sulfate attack have been revised following the publication in 2001 of BRE Special Digest 1 'Concrete in Aggressive Ground' (BRE, 2001). This document replaces BRE Digest 363 'Sulfate and Acid Resistance of Concrete in the Ground' and was written partly as a result of the relatively recent discovery of the thaumasite form of sulfate attack found in the foundations of some structures in the UK. The document is in four parts and is far more comprehensive and complex than its predecessor. It coveres all forms of aggressive ground conditions, including sulfates, acids, thaumasite and brownfield sites. Readers will need to consult this document for details of the procedures that have to be followed and because of the complexity of these procedures and the number of variables involved it is difficult to give simple examples to illustrate the recommendations resulting from guidelines. If, for example, the soil was classified as Design Class 3*(DC3) then, depending on the aggregate type and other conditions, the requirements might be as shown in Table 3.6.

Table 3.6

Cement type	Min cement (kg/m^3)	Max w/c
OPC + ≮ 70% ≯ 90% ggbs or sulfate-resisting Portland cement	380	0.45

Chloride ingress Slag cements are shown significantly more resistant to the ingress of chloride ions than Portland cements. The reason for this it is not simply due to the reduced permeability of the slag cements but also because the chlorides chemically combine with the slag hydrates which has the effect of reducing the mobility of the

---

*DC3 = 1.5 – 3.0 g/l SO_4 in groundwater which is equivalent to 1.2 – 2.5 g/l SO_3 as defined in Digest 363 and BS 8110

chlorides. This improved resistance has the potential for minimizing the risk of corrosion of the reinforcement in concrete.

Results in Figure 3.28 (Smolczyk, 1977) show that even at a high water/cement ratio of 0.7 the concrete containing slag had a significantly lower chloride diffusion rate than even the Portland cement concrete with the lowest water/cement ratio of 0.5.

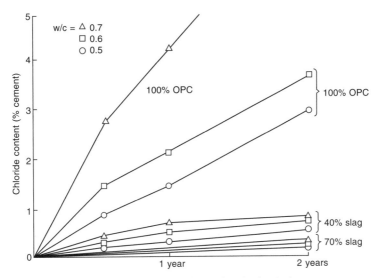

Figure 3.28 Influence of slag content on chloride diffusion (Smolczyk, 1977).

Results of a more recent long-term study using both laboratory specimens and specimens taken from real structures are summarized in Figure 3.29 (Bamforth, 1993). These data show that there is a significant reduction with time in the chloride diffusion coefficients of concretes containing ggbs whereas with Portland cement concretes there is very little change.

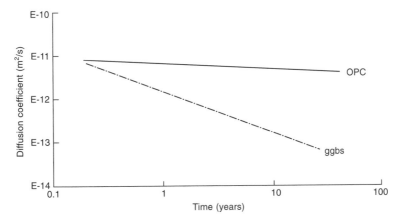

Figure 3.29 Change of effective chloride diffusion coefficient with time for concretes with and without ggbs (Bamforth, 1993).

Alkalinity   Electrochemical measurements have shown that despite the reduction in calcium hydroxide, caused by the secondary reactions of the ggbs hydration, the pH of the paste remains at a level which is well in excess of that which would affect the passivity of the reinforcing steel.

Freeze–thaw resistance   There is little difference between the freeze–thaw resistance of Portland cement and ggbs concretes of a similar strength and air content. However, non-air-entrained concretes with high slag levels (> 60 per cent) are likely to show an inferior resistance to attack than equivalent Portland cement concretes. The use of ggbs also does not normally impair the effectiveness of air-entraining admixtures to entrain air.

Abrasion resistance   Provided the concrete has been adequately cured and when comparisons are made on the basis of equal grade there would appear to be a slight advantage in terms of abrasion resistance from the use of ggbs. However, under conditions of inadequate curing the abrasion resistance of all concretes will be significantly reduced although ggbs concretes appear to be more affected than those made from Portland cement.

3.5.7 Concluding remarks

Slag concretes have been used successfully in concrete for many yeras in many countries throughout the world. There is much evidence, both from laboratory test data and in service records, to suggest that there are potentially many technical benefits to be gained from using this material. Many countries have accepted the benefits and have recommended its use in their national standards.

Provided the user is made aware of the properties of the material and has therefore an understanding of both the advantages and disadvantages to be gained, there is no reason why it should not continue to be used successfully and more often in the future.

3.6 Silica fume for concrete

3.6.1 The material

The terms condensed silica fume, microsilica, silica fume and volatilized silica are often used to describe the by-products extracted from the exhaust gases of silicon, ferrosilicon and other metal alloy smelting furnaces. However, the terms microsilica and silica fume are used to describe those condensed silica fumes that are of high quality, for use in the cement and concrete industry. The latter term, silica fume, will be the one used in the new European Standard – prEN 13263-1.

Silica fume was first 'obtained' in Norway, in 1947, when environmental restraints made the filtering of the exhaust gases from the furnaces compulsory. The major portion of these fumes was a very fine powder composed of a high percentage of silicon dioxide. As the pozzolanic reactivity for silicon dioxide was well known, extensive research was undertaken, principally at the Norwegian Institute of Technology. There are over 3000 papers now available that detail work on silica fume and silica fume concrete.

Large-scale filtering of gases began in the 1970s and the first standard, NS 3050, for use in a factory-produced cement, was granted in 1976.

Production and extraction

Silica fume is produced during the high-temperature reduction of quartz in an electric arc furnace where the main product is silicon or ferrosilicon. Due to the vast amounts of electricity needed, these furnances are located in countries with abundant electrical capacity including Scandinavia, Europe, the USA, Canada, South Africa and Australia.

High-purity quartz is heated to 2000°C with coal, coke or wood chips as fuel and an electric arc introduced to separate out the metal. As the quartz is reduced it releases silicon oxide vapour. This mixes with oxygen in the upper parts of the furnace where it oxidizes and condenses into microspheres of amorphous silicon dioxide.

The fumes are drawn out of the furnace through a precollector and a cyclone, which remove the larger coarse particles of unburnt wood or carbon, and then blown into a series of special filter bags.

The chemistry of the process is very complex and temperature dependent. The SiC formed initially plays an important intermediate role, as does the unstable SiO gas which forms the silica fume:

At temperatures > 1520°C
$$SiO_2 + 3C = SiC + 2CO$$
$\downarrow$
$\downarrow$

At temperatures > 1800°C
$$3SiO_2 + 2SiC = Si + 4SiO + 2CO$$

The unstable gas travels up in the furnace where it reacts with oxygen to give the silicon dioxide:

$$4SiO + 2O_2 = 4SiO_2$$

The silica fume is not collected as pure material as other particles and chemicals are present in the powder collected in the filter bags (See Figures 3.30 and 3.31)

Characteristics

Silica fume is, when collected, an ultrafine powder having the following basic properties:

1 At least 85 per cent SiO_2 content.
2 Mean particle size between 0.1 and 0.2 micron.
3 Minimum specific surface of 15 000 m^2/kg.
4 Spherical particle shape.

The powder is normally grey in colour but this can vary according to the source. Operation of the furance, the raw materials and quality of metal produced, will all have an effect on the colour of the powder. This variation can show up in the material from one furnace as well as from different furnaces, thus it is advisable to ensure a consistency of supply from one source (Table 3.7).

Health and safety

As with all fine powders there are potential health risks, particularly in relation to silicon dioxide and the lung disease silicosis. In studies of the material and of workers in the ferrosilicon industry (Aitcin and Regourd, 1985; Jorgen, 1980) it has been found that the

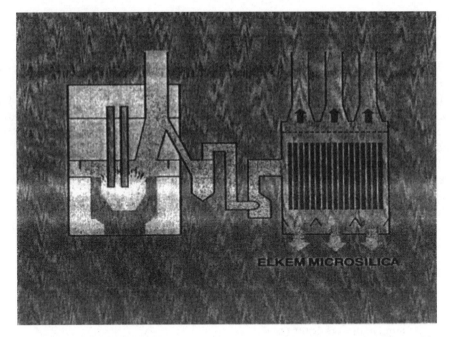

Figure 3.30 Typical plant layout: (left to right): furnace, main stack, pre-collector, cooling pipes and fan, and baghouse. (Courtesy Elkem Materials Ltd).

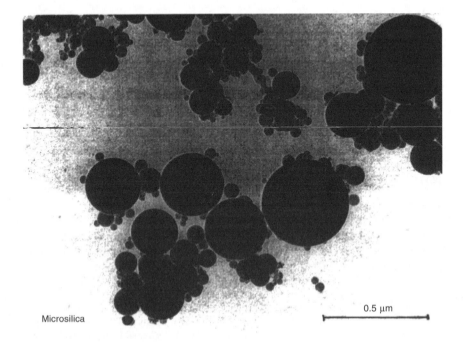

Figure 3.31 Undensified silica fume (SEM).

Table 3.7 A comparison of cementitious materials available in the UK

	Portland cement	pfa	ggbfs	Microsilica
Physical data for cementitious materials				
Surface area (m²/kg)	350–500	300–600	300–500	15 000–20 000
Bulk density (kg/m³)	1300–1400	1000	1000–1200	200–300
Specific gravity	3.12	2.30	2.90	2.20
Chemical data for cementitious materials				
SiO_2	20	50	38	92
Fe_2O_3	3.5	10.4	0.3	1.2
Al_2O_3	5.0	28	11	0.7
CaO	65	3	40	0.2
MgO	0.1	2	7.5	0.2
$Na_2O + K_2O$	0.8	3.2	1.2	2.0

silicon dioxide that causes silicosis is the crystalline form. Silica fume, produced as described above, is amorphous and has been found to be non-hazardous.

The threshold limit values for the respirable dust have been set at 3 mg/m³ (low) and 5 mg/m³ (high) and exposure to concentrations above the high limit are not advised. The CAS number for both powder and slurry is:

$$\text{Amorphous } SiO_2: 7631\text{-}86\text{-}9.$$

For full health and safety information it is always advised to contact the producers.

Available forms of silica fume

As the powder is a hundred times finer than ordinary Portland cement there are transportation, storage and dispensing considerations to be taken into account. To accommodate some of these difficulties the material is commercially available in various forms. The differences between these forms are related to the shape and size of the particles and do not greatly affect the chemical make-up or reaction of the material. These differences will influence the areas of use and careful thought should be given to the type of silica fume chosen for a specific application. The main forms are as follows.

Undensified

$$\text{Bulk density: } 200\text{–}350 \text{ kg/m}^3$$

Due to the very low bulk density, this form is considered impractical to use in normal concrete production. The main areas of use are in refractory products and formulated bagged materials such as grouts, mortars, concrete repair systems and protective coatings.

Densified

$$\text{Bulk density: } 500\text{–}650 \text{ kg/m}^3$$

In the densification process the ultrafine particles become loosely agglomerated, making the particulate size larger. This makes the powder easier to handle, with less dust, than the undensified form. Areas where this material is used are in those processes that utilize high shear mixing facilities such as pre-cast works, concrete rooftile works or readymixed concrete plants with 'wet' mixing units.

Micropelletized

Bulk density: 700–1000 kg/m^3

Micropelletization involves forming the powder into small spheres about 0.5–1 mm in diameter. The material in this form does not readily break down in conventional concrete mixing and is best suited to intergrinding with cement clinker to produce a composite cement. Icelandic cement is made in this fashion and contains 7.5 per cent silica fume to combat the potential ASR from the local aggregates.

Slurry

Specific gravity: 1400 kg/m^3

This material is produced by mixing the undensified powder and water in equal proportions, by weight, to produce a stable slurry.

Mixing and maintaining a stable slurry requires expensive hi-tech equipment and cannot be done easily and, therefore, all slurries shold be obtained from one specific supplier to maintain quality. In this form the material is easily introduced into the concrete mix. It is also the most practical form for dispensing by weight or volume. It can be used for virtually all forms of concrete from semi-dry mixes for pre-cast products to self-compacting concrete and is ideally suited to the readymixed industry.

3.6.2 Inclusion in concrete

Standards and specifications

Various national standards, codes of practice and recommendations have been published since the use of silica fume has become globally accepted (See Table 3.8). The current list is as follows:

- Arabian Gulf – Rasheeduzzafar *et al*. Proposal for a code of practice to ensure durability of concrete construction in the Arabian Gulf environment. King Faud University, Saudi Arabia.
- Australia – AS 3582.3 1994
- Canada – CAN/CSA-A23.5-M98. Supplementary cementing materials.
- China – draft
- Denmark – DS 411. Code of practice for the structural use of concrete.
  The Danish Academy of Technical Sciences Hefte no. 25.
  Basic Concrete Specification. Copenhagen, May 1986
- Finland – Suomen Betoniyhdistys r.y. Betoninormit 1987, RakMK BY, by 15, Helsinki
- Germany – Institut Für Bautechnik. PA VII-21/303
- India – draft (IS 1727)
- Japan – JIS A6207:2000
- New Zealand
- Norway – NS 3098. Portland cements, specification of properties, sampling and delivery.
  NS 3420. Specification texts for building and construction, NBR, Oslo
- Sweden – Statens Planverk PFS 1985:2. Mineral additions to concrete, approval rules.
  Staten Planverk, Stockholm

Table 3.8 Standards comparison table

	USA ASTM C1240:01	Norway NS 3045	Canada CSA, A23.5-98	Australia AS 3582	Europe CEN prEN 13263-1 (Provisional)	France NF P 18-502 1992	Japan JIS A6207: 2000
SiO_2 %>	85.0	85	85	85	85	85	85
SO_3 %<			1.0	3	2.0	2.5	3.0
Cl %<		Report if >0.10		Report	0.3	0.2	0.1
CaO %<		2			1.0	1.2	1.0
MgO							5.0
Si (free) %<					0.4		0.4
Total alkalis						4	
Free C						4	
Moisture content %<	3.0			2		1	3.0
LOI %<	6.0	5	6.0	6	4.0		5.0
Specific surface m²/g >		>12			15–35	20–35	>15
Bulk density	Report			Report			
Pozzolanic activity index, %>	85% – 7d accel'd curing, w/cm = variable	95% – 28d Normal curing w/c = 0.5	85% – 7d accel'd curing	Report	100% 28d Normal curing w/p ratio = 0.5		95% – 7d 105% – 28d w/c = 0.5
Retained on 45 micron sieve % <	10		10				
Density, kg/m³						2100–2300	Report
Autoclave expansion % <			0.2%				
Canadian Foaming test			No visible foam				
Notes		Characteristic values			Characteristic values. Not an official standard, in the approval process.	Material type A	

Note: The main standards that are accepted on a global basis are the ATSM C1240 and the CSA A23.5. Only the latest versions of any of these standards should be accepted for specification use. The table gives only the mandatory chemical and physical requirements. Several of the standards also contain optional requirements. Where there are blanks, no mandatory requirements exist.

- UK – Agrément Certificate No. 85/1568. British Board of Agrément
- USA – ASTM C-1240.xx. (always use newest version)
- EU – prEN 13263-1 (expected vote 2003)

Effects on fresh concrete

Due to the nature and size of the silica fume, a small addition to a concrete mix will produce marked changes in both the physical and chemical properties. The primary physical effect is that of adding, at the typical dosage of 8–10 per cent by cement weight, between 50 000 and 100 000 microspheres per cement particle. This means that the mix will be suffused with fine material causing an increase in the cohesiveness of the concrete. When using a powder form of silica fume this will mean an increased water demand to maintain mixing and workability, and therefore powders are most often used with plasticizers or superplasticizers.

Regarding workability, it should be noted that a fresh silica fume concrete will have a lower slump than a similar ordinary concrete due to the greater cohesion. When the mix is supplied with energy, as in pumping, vibrating or tamping, the silica fume particles, being spherical, will act as ball bearings and lubricate the mix giving it a greater mobility than the similar ordinary concrete. Silica fume concrete is often referred to as being thixotropic in nature to describe this. Thus when measuring the slump of a silica fume concrete it must be remembered that the value will only indicate the consistency of the concrete and will not relate to its workability. The most favourable test for such concrete is the DIN Standard flow table, or similar, which gives a reaction to an energy input and thus gives a better visual appraisal of the workability of the mix.

In mixes using the slurry material as an addition there can be a slight increase in water demand to maintain a given slump. This demand is normally offset by the use of a standard plasticizer. Another way of negating the effect is by reducing the sand content.

In the readymix industry use of the slurry material is nearly always enhanced by a nominal dosage of a plasticizer to ensure full dispersion and maintain the w/c ratio. As the concrete is more cohesive it is less susceptible to segregation, even at very high workabilities such as in flowing or self-compacting concretes. This lack of segregation also makes it ideal for incorporation into high-fluidity grout.

The ultrafine nature of the particles will provide a much greater contact surface area between the fresh concrete and the substrate or reinforcement and thus will improve the bond between these and the hardened concrete.

Another aspect of this non-segregation, and the filling of the major voids in the fresh concrete, is that a silica fume concrete will produce virtually no bleedwater. The concrete must therefore be cured, in accordance with good site practice, as soon as it has been placed, compacted and finished. The lack of bleedwater means that processes such as powerfloating can be commenced much sooner than with ordinary concretes.

However, this lower slump, lack of bleedwater and 'gelling' (stiffening when not agitated) does not indicate a rapid set. Silica fume is a pozzolana and, therefore, requires the presence of calcium hydroxide to activate it. The calcium hydroxide is produced by the cement hydrating and thus the silica fume can only be activated after the cement has started reacting. The setting times for microsilica concretes should be similar to those of ordinary concretes except when specifically designed for such features as ultra-high strength or low heat.

Mix design criteria

It is considered by most of the producers of silica fume that the powder forms are best suited to specialized production or mixing facilities or precasting and readymix production using high power forced action mixers. This is due to the water demand of the powders and the subsequent difficulty in dispersion in a drier mix.

In most readymix or precast operations it is normal for the slurry product to be introduced at the same time as the mix water. Although the slurry disperses more readily, care is needed to ensure a uniform quality of mix before allowing the concrete to be used.

There are variations of dosage for given types of application which serve as guidelines for initial trial mixes. It should be stressed that, even with the precedent of past work, trial mixes should always be conducted before acceptance of a mix formulation. The values in Table 3.9 are often used as starting points:

Table 3.9

Type of concrete	Dosage as % of total cementitious content
Normal	4–7
High performance	8–10*
High chemical resistance	10–12*
Underwater	10–15
Pumped	2–5

*Higher values may be used in special circumstances.

The fine filler effect caused by adding a material with a bulk density one fifth of the cement will nearly always require a modification to the mix constituent proportions, particularly the coarse/fine aggregate ratio, to achieve optimum rheology. In nearly all readymix production a plasticizer – or, more often, a superplasticizer – is used to give optimum dispersion, while maintaining the w/c ratio.

In some countries microsilica is added as a large-scale cement replacement in readymix production but this practice is decreasing as quality control restrictions become tighter.

Mixing times will need to be adjusted to allow for maximum dispersion of the silica fume. This is most important when using any of the powder forms to prevent agglomerations within the mix. The slurry form, when used in readymixed concrete, will not need excessive mixing times since the 'mass action' of a truck mixer is known to produce better results than laboratory trials for the same mix.

As with most cementitious materials, silica fume will function more efficiently with some types of chemical admixtures than with others and trial work is advised. Most of the producers of silica fume will have basic mix designs for various situations and aggregate types.

Effects on setting and hardening concrete

As the concrete sets and hardens the pozzolanic action of the silica fume takes over from the physical effects. The silica fume reacts with the liberated calcium hydroxide to produce calcium silicate and aluminate hydrates. These both increase the strength and reduce the permeability by densifying the matrix of the concrete.

Silica fume, having a greater surface area and higher silicon dioxide content, has been found to be much more reactive than pfa or ggbs (Regourd, 1983). This increased reactivity appears to increase the rate of hydration of the C_3S fraction of the cement in the first instance (Andrija, 1986), thus creating more calcium hydroxide, but settles down to more normal rates beyond two days.

The high reactivity and consumption of calcium hydroxide has prompted questions relating to the pH level of the concrete and the corresponding effects on steel passivity and carbonation rate. Studies have shown that the effect on carbonation rate is highly dependent on the quality of the concrete mix produced. Good-quality, well-proportioned, silica fume concrete does not exhibit any greater carbonation than a normal Portland cement concrete. The reduction of pH in a concrete mix is usually from approximately 13.5 to 13.0 and this latter value is well above the level for steel passivity. It has been estimated that a 25 per cent addition of silica fume (Andrija, 1985) would be required to use up all the calcium hydroxide produced in a concrete, and studies have shown that at this level the pH still does not drop below 12.0. In normal practice the highest dosage advised for concrete is 15 per cent and this should have no deleterious effects.

This reduction in alkalinity and the binding of the K^+ and Na^+ ions in the pore solution (Page and Vennesland, 1983) is one of the ways in which the addition of silica fume decreases the risk of ASR in concrete (Parker, 1986a).

As the microsilica reacts, and produces the calcium silicate hydrates, the voids and pores within the concrete are filled as the crystals formed bridge the gaps between cement grains and aggregate particles. Coupling this with the physical filling effect it can be seen that the matrix of the concrete will be very homogenous and dense, giving improved strength and impermeability (Diamond, 1986). It has been found that the relatively porous section that surrounds the aggregate grains in normal concrete is virtually absent in high-quality silica fume concrete.

The cementing efficiency, a measure of reactivity, of microsilica has been found to be between four and five times that of ordinary Portland cement (Page, 1983). This implies that large amounts of Portland cement could be replaced by small dosages of silica fume and a concrete would still achieve the required strength. While this is possible, it is not considered to be an 'ethical' usage. Reducing the cementitious content to very low levels, though still achieving strength, will have adverse effects on the durability of the concrete despite the benefits imparted by incorporating the silica fume.

Even though the rate of reaction is very high in the initial stages, not all the silica fume is used up and studies have shown that unreacted material is still present at later stages (Li et al., 1985).

In general, the heat evolution of a silica fume concrete depends on the mix design. If a high early strength is needed, addition of silica fume to a high cement content mix will produce a higher heat of hydration for the initial stages, lessening as the reactivity slows. Additions to 'normal' concretes do not usually produce significant changes in heat evolution.

Silica fume concrete is very susceptible to temperature variations during the hardening process. The rate of strength gain can be reduced at temperatures, below the 20°C optimum and accelerated with increased temperatures (Sandvik, 1981). This relates to concrete and not to some of the specialized repair materials available which can be used at very low temperatures.

Blended cement mixes

While silica fume is compatible with both pulverized fuel ash and ground granulated blastfurnace slag, it is a pozzolanic material and hence will give differing results depending on the mix designs used. If present in high proportions the reactivity of pfa will be affected by the ability of the microsilica to rapidly consume the calcium hydroxide. This may provide high early strengths but a reduced rate of long-term gain. The ggbs blends are less affected because of the latent hydraulic nature of the ggbs. When high replacement levels of ggbs are used (say 50–70 per cent) silica fume can be added to improve the early age strength, as it reacts faster in the first three days, or to improve the consistency of the fresh concrete. High levels of ggbs can cause problems with high water contents leading to segregation and bleeding, not only on the surface of the concrete but also within the matrix itself. The silica fume will virtually eliminate this bleeding and hence maintain the integrity of the concrete. With more normal levels of pfa and ggbs (say 25–40 per cent) silica fume can be added to give enhanced performance. In such cases, where there would be a minor reduction in strength due to using these additions, this is offset by the silica fume and high early, and ultimate, strength can be achieved without an excessive increase in the cost of the concrete. These triple blend cements exploit the beneficial characteristics of both pozzolanic materials in producing a durable concrete. This type of concrete is being specified where concrete structures are expected to last for upwards of 100 years such as the Storebaelt in Denmark and the Tsing Ma bridge in Hong Kong. Here again, caution must be exercised and full trials conducted. While silica fume can enhance concretes with high replacement levels of pfa or ggbs, none of them can react properly without sufficient cement in the mix to produce calcium hydroxide.

There will always be a point of no return in replacing cement with pozzolanas and the target should always be the ultimate quality of the concrete not just the required compressive strength.

3.6.3 Hardened concrete: mechanical properties

Compressive strength

High compressive strength is generally the first property associated with silica fume concrete. Many reports are available (Loland, 1983; Loland and Hustad, 1981; Sellevold and Radjy, 1983) showing that the addition of silica fume to a concrete mix will increase the strength of that mix by between 30 per cent and 100 per cent dependent on the type of mix, type of cement, amount of silica fume, use of plasticizers, aggregate types and curing regimes. In general, for an equal stength, an increase will be seen in the w/c ratio while for a given w/c ratio an increased strength will result (Sellevold and Radjy, 1983) (Figure 3.32). Silica fume concrete will show a marked reduction in strength gain when subjected to early age drying (Johansen, 1981) and this can be as much as 20 per cent. Combinations of silica fume and pfa do not seem as subject to these changes (Maage and Hammer, 1985).

With correct design, concrete with ultra-high strength can be produced using normal readymix facilities. In the USA it is common to use 100–130 MPa concrete in tall buildings, and 83 MPa silica fume concrete was used to build the 79-storey office block at 311 South Wacker in Chicago. 100 MPa silica fume concrete was used in the Petronas Towers in Kuala Lumpur. (See also Chapter 1 on High Performance Concrete)

Tensile and flexural strength

The relationships between tensile, flexural and compressive strengths in silica fume concrete are similar to those for ordinary concrete. An increase in the compressive strength using silica fume will result in a similar relative increase in the tensile and flexural strength. This plays a strong role when silica fume concrete is used in flooring, bridging or roadway projects. The increased tensile strength allows for a possible reduction in slab thickness, maintaining high compressive strengths, thus reducing overall slab weight and cost.

Curing has a large influence on these particular properties and it has been shown that poor curing at early ages will significantly reduce the tensile and flexural strength (Johansen, 1981).

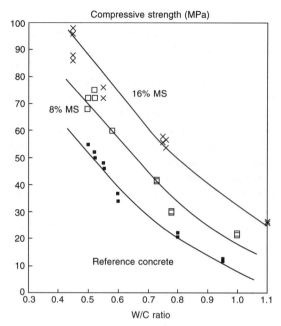

Figure 3.32 Strength versus w/c ratio.

Brittleness and E modulus

The stronger a concrete is the more brittle it becomes and silica fume concrete is no exception to this rule. However, the strengths concerned are the high to ultra-high values, 150–250 MPa and over, and not the 50–100 MPa concrete being used for most projects (Sellevold, 1982a,b; Justesen, 1981).

E modulus does not follow the pattern of tensile strength and only shows slight increase in comparison to compressive strength. Therefore high- and ultra-high strength concretes can be used for tall structures without loss of ductility (Larrard, 1987).

Bonding

Silica fume concrete has a much finer paste phase and the bond to substrates, old concrete, reinforcement, fibres and aggregates will be improved. Investigation has shown (Carles-Gibergues, 1982) that the aggregate–cement interface is altered when silica fume is present, and pull-out tests (Gjorv et al., 1986; Monteiro et al., 1986) show improved strength. Bonding to fibres is greatly improved (Bache, 1981; Krenchel and Shah, 1985; Ramakrishnan and Srinivasan, 1983). This is particularly beneficial in the steel fibre/silica fume modified shotcrete which is widely used in Scandinavia. It negates the use of mesh reinforcement which would otherwise have to be fixed to the substrate. The high reactivity and extreme cohesiveness of silica fume shotcrete or gunite also reduces the need for an accelerator. Such use of silica fume has resulted in rebound figures of less than 5 per cent.

Shrinkage

Shrinkage in cement pastes has been found to be increased when using silica fume and so when paste, mortar or grout systems are used, a compensator is usually added.

In concretes the shrinkage is related to the aggregate volume and aggregate quality and many reports are available (Traettenberg and Alstad, 1981; Johansen, 1979), that show that the addition of silica fume will reduce shrinkage in concrete when close control is exercised over mix design and aggregate selection. The importance of good curing is again stressed in these papers.

Creep

There is little information about the effect of silica fume on the creep of concrete. What is available refers to high-strength, high-dosage mixes and, in general, for such concretes it appears that silica fume concrete exhibits less creep compared to ordinary concrete (Johansen, 1979; Buil and Acker, 1985).

Fire resistance

It is known that high-strength concretes may explode when exposed to fire. Several tests (Wolsiefer, 1982; Maage and Rueslatten, 1987; Shirley *et al.*) have shown that under normal fire conditions silica fume concrete behaves in a similar way to normal concretes. Ultra-high-strength silica fume concrete may be susceptible to this type of failure due to increased brittleness and consideration should be given to the mix design using low free water content.

Abrasion and erosion

Low w/c ratio high-strength silica fume concrete shows greatly improved resistance to abrasion and erosion and a large amount of silica fume concrete has been produced to specifically utilize this quality. A large repair project on the Kinzua dam, USA, has been studied (Holland, 1983) and results show good performance of the concrete used. Many hydropower projects in India are utilizing silica fume concrete for this performance.

3.6.4 Hardened concrete: durability-related properties

The use of a silica fume concrete, with its potential for greater strengths, both compressive and tensile, its more refined pore structure and lower permeability, gives the opportunity of providing a more durable concrete with a longer working lifespan than a conventional concrete in the same environment.

Permeability

The two main methodologies of measuring permeability are either statically, such as allowing a concrete to dry out and noting the weight loss, or actively, by subjecting the material to a liquid or gas under pressure and measuring the depth of penetration. In studies using drying methods (Sellevold *et al.*, 1982a, b; Sorensen, 1982) the efficiency factor for silica fume concrete was between 6 and 8. This indicates that the physical size and high reactivity of the silica fume have more influence on permeability than on compressive strength.

In active testing, the permeability under water pressure, early tests (Sorensen, 1982) showed results ranging from very little permeability to impermeable. Since these tests (carried out in the 1960s) several comprehensive evaluations have been made (Markestad, 1977; Hustad and Loland, 1981; Sandvik, 1983) which confirm the previous results. Examinations on mature specimens (Skurdal, 1982) gave reduced permeability and microscopic studies revealed a very dense microstructure and the virtual absence of the weak layer normally surrounding the aggregate grains.

In all these tests close attention was paid to the curing regimes used and it was found that the curing time had a very marked influence on the results. Permeability should not be confused with porosity, as the pore structure is modified but not decreased (Sellevold *et al.*, 1982a, b).

The permeability of concretes, particularly to chemicals such as chlorides and sulfates, is a great concern around the world with regard to durability. Research has been ongoing and results are frequently added to the list of references.

Sulfate resistance

In a major study initiated in Oslo in the first years of testing silica fume concretes various specimens were buried in the acidic, sulfate-rich ground in Oslo. The 20-year results are available for this trial (Maage, 1984; Fiskaa, 1971) and indicate that the silica fume concretes performed as well as those made with sulfate-resisting cement. This is confirmed in the 40-year report and in further laboratory tests (Fiskaa, 1973, Mather, 1980).

When testing in conjunction with ggbs or pfa (Carlsen and Vennesland, 1982), silica fume mixes were found to show greater resistance to sulfate attack than those made with special sulfate-resisting cements. This has resulted in silica fume concrete being specified in areas such as the Arabian Gulf to combat severe deterioration in the concrete (Rasheeduzzafar *et al.*).

Such performance of the concrete can be attributed to:

- The refined pore structure and thus the reduced passage of harmful ions (Popovic, 1984).
- The increased amount of aluminium incorporated into the microsilica, thus reducing the amount of alumina available for ettringite formation.
- The lower calcium hydroxide content.

Chloride resistance

Chlorides can penetrate concrete from external sources or be present in the mix constituents. The ability of concrete to withstand chloride penetration from, for example, sea water or de-icing salts is important, as is the ability of the mix to bind the aggressive fraction of the chlorides present in the pore water. Studies have been made (Page and Vennesland, 1983; Mehta, 1981; Page and Havdahl, 1985; Monteiro *et al.*, 1985) which show the varying effects of the lower permeability and the reduction of pH in the pore water and how this relates to the presence of chlorides and the state of passivity of embedded steel.

It is considered that the lower the pH, then the lower the threshold limit for depassivation of the steel by chlorides. In studies of the penetration coefficient of chlorides (Fisher, 1982) this was shown to be much lower in silica fume concretes, thus negating the effect of the lower threshold value.

In general for equivalent strengths, initiation of chloride attack will be delayed in a concrete containing silica fume.

Providing sufficient cover to reinforcement is also important since reducing the penetration rate is ineffective if the cover is also reduced.

In the USA a special rapid test – ASTM 1202 (Christensen *et al.*, 1984) – has been applied to compare the chloride penetration of high-strength silica fume concrete with latex modified and low w/c ratio concrete. The test is based on the resistivity of the concrete to the passage of an electrical charge. Properly designed, produced and finished silica fume concrete normally achieves a very low rating.

Carbonation

Results for tests on carbonation rates are somewhat varied and contradictory depending on the viewpoint taken when analysing the findings. A study (Vennesland and Gjorv, 1983) into the effect of microsilica on carbonation and the transport of oxygen showed that adding up to 20 per cent microsilica caused a slight reduction in these two actions in water-saturated concrete. In essence, the conclusions shown by those reports available (Johansen, 1981; Vennesland and Gjorv, 1983; Vennesland, 1981) are that for equal strengths and any concretes below 40 MPa carbonation is higher in silica fume concrete. Concretes above 40 MPa show reductions in carbonation rate and it is only these concretes that are deemed susceptible to attack and damage if there is reinforcement present.

As silica fume concrete is normally used where the compressive strengths are above 40 MPa it is a moot point as to whether carbonation is a serious risk. Correct curing procedures are essential to ensure optimum performance of the silica fume concrete.

Efflorescence

These actions occur mainly when one or more surfaces of the concrete is subjected to either continuous water contact or intermittent wetting and drying. The excess calcium hydroxide is leached through the concrete to the surface where it carbonates, giving a white powdery deposit. Efflorescence not only reduces the aesthetic quality of structures but can also result in increased porosity and permeability and ultimately a weaker and less durable concrete. It has been found in studies (Samuelsson, 1982) that the addition of silica fume will reduce efflorescence due to the refined pore structure and increased consumption of the calcium hydroxide. The results indicated that the more efficient the curing and the longer curing time before exposure, the more resistant the concrete became.

Frost resistance

It will be appreciated that as the major producers and users in the early years were the Scandinavian countries this particular property of silica fume concrete has been well scrutinized. Investigations have included using silica fume as an addition on its own and with superplasticizers, varying dosages of air entrainers, different aggregates and various curing regimes.

For air-entrained conctete of the required strength it is necessary to achieve the correct amount of air, the right dispersion of the bubbles and a mix stable enough under compaction.

Many different concretes have been compared (Okkenhaug and Gjorv, 1982; Okkenhaug, 1983) to determine the effect of silica fume addition. It was found to be difficult to entrain air in a silica fume mix that did not use a plasticizer but that by increasing the dosage of air entrainer and adding a plasticizer it was easy to achieve the desired levels. There is some speculation as to the reason for the increased dosage of air entrainer with the most likely one being that the air and the silica fume compete for the same space in the mix. Once in the mix it was noted that the bubble spacing and stability were greatly improved. The variations of air content for given dosages, as sometimes happens when using pfa, were not noticed in the silica fume concretes.

The use of silica fume with air entrainment is considered to be the best option with the microsilica maintaining good stability and uniform bubble spacing of the air which gives maximum frost protection, based around a mix design guideline for a concrete of 30–50 MPa, utilizing 8 per cent microsilica and 5 per cent air entrainment as an optimum. In all cases the concrete should be cured for the longest allowable time before exposure to the working environment.

Alkali–silica reaction

To view the effect of silica fume on this form of attack it is necessary to remember the three main factors that are required for potential reaction.

- A high alkali content in the mix
- Reactive aggregates
- Available water

Silica fume reacts with the liberated calcium hydroxide to form calcium silicate hydrates and this reduction leads to a lowering of the pH and a lower risk of reaction due to high alkalis. In the formation of the calcium silicate hydrates the K^+ and Na^+ ions are bound in the matrix and cannot react with any potentially siliceous aggregates.

The minute size and pozzolanic reactivity of the silica fume greatly refines the pore structure of the concrete and reduces the permeability such that less water can enter. The normal dosage of microsilica, i.e. 10 per cent by weight of cement, can negate the main factors that could lead to alkali–silica reaction and many reports are available (Asgeirsson and Gudmundsson, 1979; Parker, 1986a, b; Perry and Gillott, 1985) that confirm this.

3.6.5 Concluding summary

The inclusion of silica fume in concrete causes significant changes in the structure of the matrix, through both physical action and a pozzolanic reaction, to produce a densified, refined pore system and greater strength. In most cases it is the refinement of the pore system which reduces penetrability, that has the greater effect on the performance of the concrete than the increased strength.

Use can be made of these improved qualities in designing concretes to comply with requirements or greater resistance to certain hostile environments.

Silica fume should be considered as an addition to a mix rather than a replacement for cementitious content and sensible mix design is essential.

Silica fume concrete is susceptible to poor curing and the effects are more pronounced than in ordinary concrete. Close attention to curing methods and times is important to ensure optimum performance.

Designing a silica fume concrete for specific requirements should always be a matter of consultation between the client, contractor, readymix or precast supplier and the silica fume supplier. Reference should be made to any previous project or use and trial work is essential to ensure correct use of the material and to allow an appreciation of the characteristics of this type of concrete.

3.7 Metakaolin

3.7.1 Occurrence and extraction

Kaolin is soft, white clay resulting from the natural decomposition of feldspars and other clay minerals. It occurs widely in nature. It is used for making porcelain and china, as a filler in the manufacture of paper and textiles and as a medicinal absorbent. Kaolinite is the principal mineral constituent of kaolin.

In the UK the main sources of kaolin that are commercially exploited are in the south-west of England. The granites that occur in Cornwall originally contained three principal minerals: feldspar, quartz and mica. Over geological time the feldspar decomposed to form kaolin. Kaolin-rich ball clays occur in Devon and Dorset.

Kaolin is extracted from the granite using high-pressure water jets. The kaolin slurry is then concentrated and refined using standard mineral processing techniques. The refined product is dried. The ball clays are treated using standard dry processing techniques.

3.7.2 Metakaolin production

When kaolin is heated to a temperature of 450°C dehydroxylation occurs and the hydrated aluminosilicates are converted to materials consisting predominantly of chemically combined aluminium, silicon and oxygen. The rate at which water of crystallization is removed increases with increasing temperature and at 600°C it proceeds to completion (ECCI, 1992; Highley, 1984). Metakaolin is formed in kilns when kaolin is heated at a temperature between 700°C and 800°C. The calcined product is cooled rapidly and ground to a fine powder. The metakaolin formed in this way has a highly disorganized structure.

3.7.3 Physical properties

The physical properties of metakaolin depend very much on the quality of the raw material used, the calcination temperature and the finishing processes. The products available in the UK have the typical properties listed below:

Fineness:
> 10 microns (mass % max.)	10
< 2 microns (mass % min.)	5
Surface area (m²/g)	12.0

Moisture:
% Water (mass % max.)	0.5
Bulk density: (g/cm³)	0.3

3.7.4 Reaction mechanisms

Because of its highly disorganized structure metakaolin reacts very rapidly with the calcium hydroxide produced during the hydration of Portland cements. The reactivity of metakaolin may be compared with that of other commonly available industrial pozzolans using the Chapelle test. In this test, a dilute slurry of the pozzolan is reacted with an excess of calcium hydroxide at 95°C for 18 hours. After this period the quantity of calcium hydroxide consumed in the reaction is calculated. Table 3.10 gives details of results of the Chapelle test reported by Largent (1978).

Table 3.10 Reactivity of pozzolans using the Chapelle test

Pozzolan	Pozzolan reactivity (mg $Ca(OH)_2$ consumed per g of pozzolan)
Ggbs	40
Microsilica	427
Pfa	875
Metakaolin	1050

More recently Larbi and Bijen (1991) showed that calcium hydroxide was virtually eliminated from a cement matrix that contained metakaolin. Jones *et al.* (1992) confirmed this, Figure 3.33 summarizes the results of their experiments. In spite of the fact that calcium hydroxide levels are reduced significantly, the pH of the pore solution is maintained above 12.5 (Asbridge *et al.*, 1992).

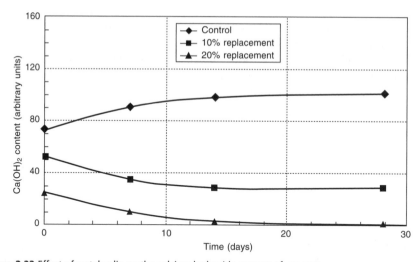

Figure 3.33 Effect of metakaolin on the calcium hydroxide content of concrete.

Several workers have studied the products of the reactions between metakaolin and calcium hydroxide. Turrizani (1964) reported that the cementitious hydrates gehlenite and tobermorite were formed by the reaction. The formation of a number of cementitious compounds (calcium silicate hydrates and a rage of calcium aluminosilicate hydrates) and the acceleration of the hydration of C_3S have been reported from studies of reactions between metakaolin and calcium hydroxide in the presence of other cement hydrates (Ambroise, 1992; Bredy *et al.*, 1989; Murat, 1983; Pietersen *et al.*, 1993). Dunster *et al.*, (1993) observed that metakaolin accelerated the early hydration of calcium silicate to produce extra polymeric silicate gel.

The formation of insoluble, stable cementitious products in place of potentially soluble calcium hydroxide not only reduces the amount available for attack by external agencies (sulfates, acids etc.), it also modifies the pore structure of the paste phase of the concrete leading to reduced porosity and permeability. In his work on the pore solution chemistry, Halliwell (1992) showed that the concentration of alkali metal ions was reduced significantly when metakaolin replaced Portland cement in concrete. Table 3.11 shows the results of tests of concrete in which 10 per cent of the cement was replaced by metakaolin.

Table 3.11 Effect of metakaolin content on concentration of alkali metal ions in pore solution

Portland cement	Alkali content (% Na$_2$O equivalent)	Reduction of Alkali metal ion content of pore water at 1 year (%)
Low alkali	0.32	37
	0.63	46
High alkali	1.15	46

3.7.5 National Standards

No National Standards exist that cover metakaolin as a cementitious addition for concrete. The product Metastar Metakaolin is covered by Agrément Certificate Number 98/3540. This certificate relates to a specific product used at replacement levels up to 20 per cent of Portland cement.

3.7.6 The effect of metakaolin on the properties of concrete

Fresh concrete

The particle density of metakaolin (2.4) is lower than that of Portland cement (3.1). Thus, when metakaolin is used as a replacement for cement the volume of cementitious material is increased. Reducing the sand content of the mix overcomes the effect of the increased volume of cementitious powder. Shirvill's (1992) work indicates that a reduction of sand content of 5 per cent is necessary when 10 per cent of cement is replaced by metakaolin.

The cohesiveness of metakaolin concrete is greater than that of plain concrete. Metakaolin concrete is easier to pump and place generally and it bleeds less than plain concrete. The greater volume of cementitious fines results in the production of sharp arrises and high-quality surface finish on cast vertical surfaces.

The greater volume of cementitious material in metakaolin concrete increases its water demand. The effect of increasing metakaolin content on the water demand of concrete is shown in Figure 3.34. The increase of water demand is readily offset by the use of a single dose of a standard low-cost plasticizer (Martin, 1998); this effect is also shown in Figure 3.34

Hardened concrete

Compressive strength   The effect of metakaolin on the compressive strength of concrete has been widely reported. Some researchers (Larbi and Bijen, 1991; Halliwell, 1992; Saad *et al.*, 1982; Collin-Fevre, 1992) report that metakaolin has no adverse effect on compressive strength. Other researchers (Andriolo and Sgaraboza, 1986; Marsh, 1992;

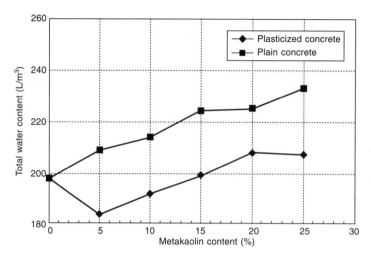

Figure 3.34 Effect of metakaolin on water demand.

Gold and Shirvill, 1992) report significant improvements of strength. This range of apparently conflicting conclusions reflects differences of factors such as:

- trial mix methodology
- approach to mix design
- variations of cement composition and
- variations in the quality of metakaolin used.

Early work by Gold and Shirvill (1992) showed that the inclusion of metakaolin in concrete produced very significant improvements in strength. The results of their work indicated that the optimum level of replacement lay somewhere between 5 per cent and 10 per cent. Increased replacement, beyond 10 per cent, did not provide further increases of strength. Sand contents were adjusted for the metakaolin concrete in these trials but no admixtures were used to offset increased water demands. Figure 3.35 illustrates the results of these early tests.

Closely monitored plant trials carried out by two RMC Readymix Companies confirmed the significant increases of strength associated with the inclusion of metakaolin found in laboratory trials.

The use of metakaolin in high strength concrete has been proposed by a number of workers (Marsh, 1992; Balogh, 1995; Calderone *et al.*, 1994). Martin (1995) reported 28-day strengths of 110 N/mm^2 for concrete containing metakaolin (10 per cent replacement of Portland cement) and a superplasticizer.

Concrete cast on site is not cured in the ideal conditions to which laboratory cured specimens are exposed. Hobbs (1996) showed that the compressive strength of plain concrete cubes and metakaolin concrete cubes was reduced when stored in laboratory air. However, the concrete containing metakaolin was stronger than the plain concrete.

Tensile strength   The flexural strength and Young's modulus of metakaolin concrete are similar to those of plain concrete of equivalent 28-day strength. It follows, therefore, that when metakaolin is used to replace a proportion of cement, flexural strength should

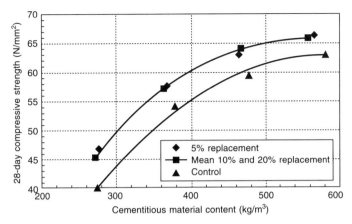

Figure 3.35 Effect of metakaolin on compressive strength of concrete.

increase. A study at Dundee University, reported by Imerys (2000), confirmed this to be the case. Table 3.12 summarizes the results of the experiments.

Table 3.12 Strength properties of concrete with a range of metakaolin contents

Mix	Cube strength (MPa)	Flexural strength (MPa)	Young's modulus (Gpa)
100% PC	41.0	5.0	30.0
90% PC, 10% Mk	47.0	Not measured	33.0
80% PC, 20% Mk	50.0	5.3	33.5

The development of tensile strength at very early ages has been studied (Hobbs, 1996). The trials showed that metakaolin significantly accelerated the development of early tensile strength.

Creep strain Creep strain has been measured for plain concrete and concrete containing metakaolin (Imerys, 2000). For cubes cured in water and then loaded to 40 per cent of their 28-day strength for 90 days, creep strain was unaffected by the inclusion of metakaolin.

Drying shrinkage The incorporation of metakaolin in concrete produces a cementitious matrix of low porosity and permeability. Water loss on drying is therefore reduced and drying shrinkage is correspondingly less. Zhang and Malhotra (1995) showed that concrete containing 10 per cent metakaolin had a lower drying shrinkage than that of plain concrete and concrete containing microsilica. Using similar test procedures Caldarone *et al.* (1994) showed that the drying shrinkage of metakaolin concrete was lower than that of plain concrete but similar to that of concrete containing microsilica.

More recently, work carried out at Dundee University (Imerys, 2000), showed that metakaolin levels up to 25 per cent did not significantly increase drying shrinkage.

3.7.7 The durability of metakaolin concrete

Durability of concrete may be defined as its ability to resist the action of weathering, chemical attack, abrasion or any other process of deterioration. In recent years this property has gained a much higher profile with engineers and specifying authorities.

With the exception of mechanical damage, many of the adverse influences on durability involve the transport of fluids through the concrete. Therefore, in general terms, high-strength low-permeability concrete should provide adequate durability. Additional considerations may be necessary in some circumstances, e.g. alkali–silica reaction. Concrete is less able to withstand some external chemical effects, e.g. acid attack. However, it is possible to provide concrete that, though affected by acids, has a significantly prolonged service life.

Resistance to freezing and thawing

The resistance of concrete to cyclic freezing and thawing depends upon a number of properties (e.g. strength of the hardened paste, extensibility and creep) but the main factors are the degree of saturation and the pore system of the hardened paste (Neville, 1995). The use of metakaolin in concrete significantly alters the pore system and permeability of the paste. It follows, therefore, that well designed metakaolin concrete will have increased resistance to the damaging effects of freezing and thawing. Zhang and Malhotra (1995) report that metakaolin concrete showed excellent performance after 300 cycles of freezing and thawing in the ASTM C666 test. The results of experiments by Calderone *et al.* (1994) confirmed that metakaolin concrete exhibited high freeze–thaw durability.

Sulfate resistance

Singh and Osborne (1994) showed that mortar prisms containing 15 per cent metakaolin showed good sulfate resistance when stored in 4.4 per cent sodium sulfate solution at 20°C. Khatib and Wild (1998) confirmed these findings. When stored at 5°C in both sodium sulfate and magnesium sulfate, sulfate resistance was further improved. Given the concerns about the formation of thaumasite at low temperatures and the subsequent deterioration of the concrete, the results of tests at 5°C are of particular interest.

Resistance to the penetration of chloride ions

The ASTM test C1202:1944 has been used to assess the ability of metakaolin concrete to resist the penetration of chloride ions. Calderone *et al.* (1994) and Zhang and Malhotra (1995) report that the resistance of metakaolin concrete to the penetration of chloride ions was significantly higher than that of plain concrete of otherwise similar composition. Ryle's work (1994) compared the results of tests using the ASTM procedure with other chloride and oxygen diffusion test methods and confirmed that concrete of very low permeability was produced when 15 per cent of the cement was replaced with metakaolin.

Table 3.13 summarizes results of tests using the ASTM procedure extracted from the literature.

Page and Coleman (1994) and Larbi and Bijen (1992) showed that the use of metakaolin as a partial replacement for cement (15 per cent and 20 per cent) significantly affected the transport processes in mortar and concrete. They showed that the partial replacement of cement with metakaolin led to a reduction of the rate of diffusion of chloride ions into concrete and mortar by an order of magnitude. Table 13.14 shows the effect of

Table 3.13 Results of chloride ion penetration tests (ASTM 1202:1994)

Research reference	Mix	Cementitious material content (kg/m^3)	Results of ASTM test (coulombs)		
			28 days	56 days	90 days
Calderone	Control	385	–	4832	–
	15% Mk	385	–	754	–
Martin	Control	385	3175	–	1875
	15% Mk	385	390	–	300
Ryle	15% Mk	400	625	395	310

Table 3.14 The effect of metakaolin on the chloride ion diffusion coefficient

Sample type	W/C Ratio	% PC Replaced with metakaolin	Chloride ion diffusion coefficient (cm^2 s^{-1}) $\times 10^{-9}$
Paste	0.5	0	90
	0.5	15	9
Mortar	0.6	0	220
	0.6	15	18
Mortar	0.7	0	680
	0.7	15	8
Mortar	–	0	597
	–	20	10

the use of metakaolin on chloride ion diffusion coefficients for mortar and paste (ECCI, 1993).

Resistance to acidic conditions

Metakaolin reacts very rapidly with the calcium hydroxide produced during hydration converting it to a variety of insoluble, stable cementitious products. In addition, by reducing the permeability of the paste and reducing the thickness of the interface between the paste and the aggregates (Larbi and Bijen, 1991, 1992; Mass, 1996) metakaolin helps to prevent the ingress of aggressive substances. It follows, therefore, that concrete containing metakaolin as a partial replacement for cement will prolong the service life of concrete subjected to attack by acidic substances.

Collin-Fevre (1992) showed that concrete containing 10 per cent metakolin as a partial replacement for cement was considerably more resistant to attack by organic and mineral acids than plain concrete of otherwise similar composition.

Dutrel and Estoup (1986) reported that concrete containing metakaolin (10 per cent replacement of Portland cement) was more resistant to the effects of lactic, formic, citric and humic acids and to mineral acids of pH 3.5.

Experiments by Martin (1997) showed that concrete containing metakaolin was more resistant than plain concrete to the acids that result from the storage of silage. To complement his laboratory work, a working silage bay was constructed using plain concrete and concrete containing metakaolin. After some 10 years in use no adverse effects are apparent on the surface of the area constructed with metakaolin concrete. Attack by silage acids had removed some of the surface of the plain concrete after 5 years of use.

Alkali–silica reaction

The suppression of alkali–silica reaction using metakaolin in concrete containing reactive combinations of aggregates is well established by laboratory experiment (Jones, *et al.*, 1992; Martin, 1993; Kostuch, 1993). Sibbick and Nixon (2000) showed that the inclusion of metakaolin in larger concrete castings stored on outdoor exposure sites prevented expansion due to alkali–silica reaction.

The earliest large-scale use of metakaolin in concrete dates back to the 1960s. Some 300 000 tonnes of locally calcined kaolin was blended with Portland cement for the construction of a series of four dams in the Amazon basin. The only aggregates available were alkali reactive. Andriolo and Sgaraboza (1986) claimed that the use of metakaolin prevented expansion due to alkali–silica reaction.

Andriolo and Sgaraboza (1986) reported additional beneficial effects resulting from the use of metakaolin. They claimed.

- substantially increased compressive strength of concrete, permitting a reduction in the amount of binder used
- reduced bleeding of the concrete
- substantially reduced temperature rise in mass concrete.

These benefits were of particular significance in view of the large volumes of concrete used in the construction of the dams.

Jones *et al.* (1992) showed that metakaolin reduced the calcium hydroxide content of the concrete to such a low level that the production of a swelling gel was not possible.

The effect of alkalis from external sources on the alkali–silica reaction has also been demonstrated (Walters and Jones, 1991; Xu and Chen, 1986). In Figure 3.36 the effect of immersing concrete prisms in sodium chloride solution is shown. Prior to immersion the prisms had been stored at a temperature of 38°C and a relative humidity close to 100 per cent to accelerate any potential reactions. By (about) 20 months the reaction in the expansive control mix had reached equilibrium. Immersion in sodium chloride solution disturbed this equilibrium and presented the control mix and the mix containing the metakaolin with additional alkalis. This caused the control mix prisms to expand rapidly again. The prisms containing the metakaolin concrete remained unaffected.

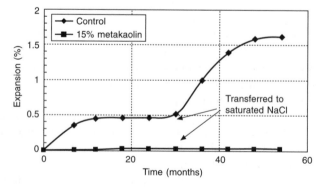

Figure 3.36 The effect of metakaolin on expansion due to ASR and immersion in NaCl.

3.7.8 Compatibility with blended cements

A large proportion of readymixed concrete manufactured in the UK contains cement replacement materials such as ground granulated blast-furnace slag (ggbs) and pulverized fuel ash (pfa). These materials are industrial pozzolans and are activated in much the same way as metakaolin.

Asbridge *et al.* (1994) showed that when used in conjunction with ggbs, metakaolin increased the strength of concrete at 96 days. The pore volume of the paste was lower than that of the corresponding control mixes containing ggbs. They considered that these changes would further improve the durability of concrete. Unpublished work by RMC Readymix (1999) shows that higher strengths and improved durability were achieved when metakaolin was used as a partial replacement for cement containing 50 per cent ggbs.

3.7.9 Efflorescence

Efflorescence is a white deposit that appears on concrete surfaces. The deposit does not affect the durability of the concrete but it detracts from the aesthetic properties of plain and coloured concrete.

There are two forms of efflorescence: primary and secondary. Primary efflorescence occurs early in the life of a structure, during the process of curing. Lime-rich water from within the concrete matrix migrates to the surface and evaporates, depositing soluble salts at the surface. The lime ($Ca(OH)_2$) reacts with atmospheric carbon dioxide to form insoluble calcium carbonate. Secondary efflorescence occurs when hardened concrete is wetted and water penetrates the surface to dissolve some of the lime remaining in the concrete. During subsequnt drying the salt-laden solution moves to the surface depositing the salts at the surface.

Metakaolin is thought to control the formation of efflorescence by:

- reducing the alkali content of the concrete, so CO_2 is absorbed less rapidly
- removing a proportion of the $Ca(OH)_2$
- refining the pore structure so the absorption of water and the diffusion of salt-laden water to the surface is reduced.

3.7.10 Summary

When kaolin is heated to a temperature in the range 700–800°C it is converted to metakaolin with a highly disorganized structure. Metakaolin is a very reactive pozzolan. In the presence of water, it reacts with calcium hydroxide to produce stable, insoluble cementitious hydrates. This pozzolanic reaction reduces the permeability and porosity of cement paste making it stronger and significantly more durable. The use of metakaolin as a partial replacement for cement in suitably designed concrete mixes has been shown to:

- improve cohesion and reduce bleeding of fresh concrete
- increase compressive strength
- reduce drying shrinkage

- improve freeze–thaw resistance
- improve sulfate resistance
- increase resistance to the penetration of chloride ions
- improve acid resistance
- eliminate alkali–silica reaction.

3.8 Limestone

3.8.1 Limestone filler

Limestone fillers are being used increasingly in cements and, more recently, as a mixer addition. EN197-1 (BS EN197-1, 2000) contains two classes of Limestone Portland cement, CEM II/A-L (or LL) and II/B-L (or LL). The former contains between 6 per cent and 20 per cent limestone and the latter 21–35 per cent limestone. This material is interground with the Portland cement clinker, which normally produces 32.5 grade cement. The requirements for the limestone are:

- The $CaCO_3$ content $\geq$ 75 per cent
- The clay content, as determined by the methylene blue test shall not exceed 1.20 g/ 100 g.
- The total Organic Carbon (TOC) shall not exceed 0.20 per cent for LL limestone or 0.50 per cent for L limestone.

The limestone filler was considered by many as inert filler, but it has been gradually accepted as contributing to the hydration process by the formation of calcium mono-carboaluminates ($C_3A \cdot CaCO_3 \cdot 11H_2O$).

In addition to factory-made Portland limestone cements, limestone fines complying with BS 7979 (2001) may also be added at the mixer to produce an equivalent combination. The specification for the filler is very similar to that for cement. Limestone fillers may also be used in special concrete, such as self-compacting concrete, to aid cohesion in the plastic concrete.

Classification of additions to BS EN206-1

Additions, that is, materials like fly ash, silica fume, limestone fillers, etc., are classified within EN206-1 (BSI, 2000) in two ways. They are defined within the standard as being 'finely divided materials used in concrete in order to improve certain properties or to achieve special properties. This standard deals with two types of additions:

- Nearly inert additions (Type I).
- Pozzolanic or latent hydraulic additions (Type II).'

The practicality of this definition is that some additions are treated as aggregate and some may be counted towards the cement content, either fully or partially. The k-value concept is applied to fly ash and silica fume within EN206-1, which are defined as Type II additions.

Smith (1967) developed a method based on applying a cementing efficiency factor known as the k factor or k-value. The mix design was adjusted as in

$$W/C_f = W/(C + k \cdot F)$$

where W/C in the equation for plain Portland cement concrete is replaced by the adjusted W/C_f ratio

W = weight of water
C = weight of Portland cement
C_f = equivalent weight of Portland cement
k = cement efficiency factor for fly ash
F = weight of fly ash

k-values can be created/applied for many purposes, e.g. equal 28-day strength, equal chloride diffusion, equal durability, etc. A calculated k-value will change depending on the Portland cement source, the curing temperature and conditions, the fly ash source, etc. The technique has been corrupted in EN206-1 that gives k-values for fly ash of 0.40 for use in combination with CEM I 42.5N or 0.20 for CEM I 32.5. These k-values are used to adjust minimum cement contents and maximum water cement ratios.

In real concretes it will be found the k-value for a fly ash, for example, will vary considerably from zero to greater than unity, depending on the materials, the temperature, the time, etc. Consequently, the k-value is one of the most variable constants known to the concrete industry.

The suitability of additions, whether they are considered to be Type I or II, the k-value or equivalent performance concepts and other factors are allowed on a National basis, providing their suitability has been established. The UK complementary standard that gives the relevant information is BS 8500 (BSI).

References

ACI (1994) *Manual of Concrete Practice, Fly ash*, 226.3R.

Aitcin, P.C. and Regourd, M. (1985) The use of condensed silica fume to control alkali silica reaction – a field case study. *Cement and Concrete Research*, **15**, 711–719.

Alasali, M.M. and Malhotra, V.M. (1991) Role of concrete incorporating high volumes of fly ash in controlling expansion due to alkali–aggregate reaction. *ACI Materials Journal*, **88**, No. 2, 159–163.

Ambroise, J., Martin-Calle, S. and Pera, J. (1992) Properties of metakaolin blended cements. *4th Int. Conf. Flyash, silica fume, slag and natural pozzolans in concrete*, Volume 1, Turkey.

Andrija D. (1986) 8th *International Congress on Chemistry of Cement*, Rio de Janeiro, Brazil. Volume 4, pp. 279–285.

Andriolo, F.R. and Sgaraboza, B.C. (1986) The use of pozzolan from calcined clays in preventing excessive expansion due to alkali–silica reaction in some Brazilian dams. *Proc. 7th Int. Conf. on AAR*. Ottawa.

Asbridge, A.H. *et al.* (1994) Ternary blended concretes-OPC/ggbs/metakaolin. *Proc. Int. Sym. Concrete across borders*. Denmark.

Asbridge, A.H., Jones, T.R. and Osborne, G.J. (1996) High performance metakaolin concrete: Results of large-scale trials in aggressive environments. *Proc. Int. Conf. Concrete in the service of mankind*, Dundee.

Asgeirsson, H. and Gudmundsson G. (1979) Pozzolanic activity of silica dust. *Cement and Concrete Research*, **9**, 249–252.

Bache, H.H. (1981) Densified cement/ultrafine particle-based materials. 2nd International Conference on Superplasticisers in Concrete, Ottawa.

Balogh, A. (1995) High reactivity metakaolin. *Concrete Construction.*

Bamforth, P.B. (1984) Heat of hydration of fly ash concrete and its effect on strength development. Ashtech '84 Conf. London, 287–294.

Bamforth, P.B. (1993) Concrete classification for R.C. structures exposed to marine and other salt laden environments. *Proceedings*, Conference on Structural Faults & Repairs. Edinburgh.

Bamforth, P.B. Taywood Engineering, England, private communication.

Bensted, J. (1988) Thaumasite – a deterioration product of hardened cement structures. *Il Cemento*, 3–10.

Bensted, J. and Barnes, P. (2001) *Structure and performance of cements*, 2nd edn, E & FN Spon, London.

Berry, E.E. and Malhotra, V.M. Fly ash in concrete. CANMET, SP85-3.

Bredy, P., Chabannet, M. and Pera, J. (1989) Microstructure and porosity of metakaolin blended cements. *Proc. Mats. Res. Soc. Sym.* Boston.

Brown, J.H. (1980) The effect of two different pulverised fuel ashes upon the workability and strength of concrete. C&CA Technical Report No. 536, June.

Browne, R.D. (1984) Ash concrete – its engineering performance. *Ash Tech* '84, London, 295–301.

BS 146. Specification for Portland-blastfurnace cement.

BS 4246. Specification for low heat Portland-blastfurnace cement.

BS 4550. Methods of testing cement.

BS 6699. Specification for ground granulated blastfurnace slag for use with Portland cement.

BS 8110. Structural use of concrete.

BS EN197-1 (2000) Cement – Part 1: Composition, specifications and conformity criteria for common cements, BSI, London.

BS EN206-1 (2000) Concrete – Part 1: Specification, performance, production and conformity, BSI, London.

BS 7979 (2001) Specification for limestone fines for use with Portland cement, BSI, London.

BS 8110. Structural use of concrete, BSI, London.

BS 8500-1. Concrete – complementary British standard to BS EN206-1, Part 1: Method of specifying and guidance for the specifier, BSI, London.

BS 8500-2. Concrete – complementary British standard to BS EN206-1, Part 2: Specification for constituent materials and concrete, BSI, London.

BS 5328: Part 2 (1997) Methods for specifying concrete, amendment 10365, May, 1999.

Buil, M. and Acker, P. (1985) Creep of silica fume concrete. *Cement and Concrete Research*, **15**, 463–466.

Building Research Establishment (1991) Sulfate and acid resistance of concrete in the ground. BRE Digest 363, July.

Building Research Establishment (1999) Alkali–silica reaction in concrete. Digest No 330. BRE, Garston.

Building Research Establishment (2001) Concrete in aggressive ground. Special Digest No. 1 2001, BRE, Garston.

Burton, M.W. (1980) The Sulphate Resistance of concretes made with ordinary Portland cement, sulphate resisting cement and ordinary Portland cement + pozzolan. Kirton Concrete Services, Humberside.

Cabrera, J.G. and Atis, C.D. (1998) Design and properties of high volume fly ash performance concrete.

Cabrera, J.G., Braim, M. and Rawcliffe, J. (1984) The use of pulverised fuel ash for construction of structural fill. Ash Tech, London.

Cabrera, J.G. and Plowman, C. (1987) Hydration and microstructure of high fly ash content concrete. Conference on concrete dams, London.

Cabrera, J.G. and Woolley, G.R. (1996) Properties of sprayed concrete containing ordinary Portland cement or fly ash Portland cement. *Proc. ACI/SCA International Conference*, Edinburgh.

Caldarone, M.A. *et al.* (1994) High reactivity metakaolin: A new generation mineral admixture. *Concrete International.*

Carette, G.G. and Malhotra, V.M. (1983) Mechanical properties, durability and drying shrinkage of Portland cement concrete incorporating silica fume. *Cement, Concrete and Aggregates*, **5**, No. 1, 3–13.

Carles-Gibergues, A. *et al.* (1982) Contact zone between cement paste and aggregate. In Bartos, P. (ed.), *Proc. International Conference on Bond in Concrete*. Applied Science Publishers, London, 24–33.

Carlsen, R. and Vennesland, O. (1982) Sementers sulfat- og sjovannsbestandighet. FCB/SINTEF, Norwegian Institute of Technology, Trondheim, Report STF65 F82010 (in Norwegian).

Christensen, D.W. *et al.* (1984) Rockbond: a new microsilica concrete bridge deck overlay material. *Proc. International Bridge Conference*, Pittsburgh, 151–160.

Christensen, P. (1982) Afprovningaf kvalitetskarakterisering af beton tilsat Silica. Teknologisk Isntitut. Byggeteknik, Copenhagen (in Danish).

Clear, C.A. and Harrison, T.A. Concrete pressure on formwork. Report 108, CIRIA, London.

Collin-Fevre, I. (1992) Use of metakaolin in the manufacture of concrete products. CIB, Montreal.

Concrete Society (1991) The use of GGBS and fly ash in concrete, Technical report 40, Crowthorne.

Concrete Society (1998) Core Project, project data for a potential revision of Technical Report No. 11.

Concrete Society (1999) Alkali silica reaction: minimising the risk of damage. Technical report No. 30, 3rd edition.

Concrete Society. Microsilica in Concrete. Technical Report 4/1993.

Davis, R.E., Carlston, R.W., Kelly, J.W. and Davis, H.E. (1937) Properties of cements and concretes containing fly ash. *ACI Journal*, **33**, 577–612.

Davis, R.E., Kelly, J.W., Troxell, G.E. and Davis, H.E. (1935) Proportions of mortars and concretes containing Portland-pozzolan cements. *ACI Journal*, **32**, 80–114.

Dewar, J.D. (1986) The particle structure of fresh concrete – a new solution to an old question. Sir Frederick Lea Memorial Lecture, Institute of Concrete Technology Annual Symposium.

Dhir, R.K. (1986) Pulverised fuel ash. *CEGB Ash-Tech 86 conference proceedings*.

Dhir, R.K., Munday, J.G.L. and Ho, N.Y. (1987) Fly ash in concrete: freeze–thaw durability. Draft report, University of Dundee.

Dhir, R.K., Munday, J.G.L. and Ong, L.T. (1981) Strength variability of OPC/fly ash concrete. *Concrete*, June.

Dhir, R.K., Munday, J.G.L. and Ong, L.T. (1986) Investigations of the engineering properties of OPC/Pulverised fuel ash concrete – deformation properties. *The Structural Engineer*, **64B**, No. 2, 36–42.

Diamond, S. (1986) 8th *International Congress on Chemistry of Cement*, Rio de Janeiro Brazil, Vol. 1, 122–147.

Dunstan, M.R.H. (1981) Rolled Concrete for Dams. CIRIA Technical note No. 106, London.

Dunster, A.M. *et al.* (1993) *Journal Mat. Sci.*

Dutrel, F. and Estoup, J.M. (1986) Effect of the addition of silica fume or pozzolanic fines on the durability of concrete products. Madrid.

ECC International. Calcined clay. Internal data sheet.

ECCI (1993) New pozzolanic materials for the concrete industry. UK.

European Standard ENV 197-1 (1992) Cement – composition, specifications and conformity criteria: Common cements.

FCB/SINTEF (1977) Norwegian Institute of Technology, Trondheim, Report STF65 A77027.

FCB/SINTEF (1984) Norwegian Institute of Technology, Trondheim, Report STF65 A84019.

Feldman, R.F. and Huang Cheng-Yi. (1984) Microstructural properties of blended cement mortars and their relation to durability. RILEM Seminar on Durability of Concrete Structures under Normal outdoor Exposure, Hanover.

FIP (1988) *State of Art Report, Condensed silica fume in conctete*, Thomas Telford, London.

Fisher, K.P. *et al.* (1982) Corrosion of steel in concrete: some fundamental aspects of concrete with added silica. Norwegian Geotechnical Institute, Oslo, Report no. 51304-06.

Fiskaa, O. *et al.* (1971) Betong i Alunskifier. Norwegian Geotechnical Institute, Oslo, Publication 86. (in Norwegian).

Fiskaa, O.M. (1973) Betong i Alkunskifier. Norwegian Geotechnical Institute, Oslo, Publication 101 (in Norwegian).

Fournier, B. and Malhotra, V.M. (1997) CANMET investigations on the effectiveness of fly ash in reducing expansion due to alkali aggregate reaction (ASR), ACAA 12th International Symposium.

Fulton, A.A. and Marshall, W.T. (1956) The use of fly ash and similar materials in concrete. *Proc. Inst. Civ. Engrs.* Part 1, Vol. 5, 714–730.

Gifford, P.M. and Ward, M.A. (1982) Results of laboratory test on lean mass concrete utilising fly ash to a high level of cement replacement. *Proc. International Symposium*, Leeds, 221–229.

Gjorv, O.E. *et al.* (1986) Effect of condensed silica fume on the steel–concrete bond. Norwegian Institute of Technology, Trondheim, Report BML 86.201.

Glasser, F.P. and Marr, J. (1984) The effect of mineral additives on the composition of cement pore fluids. *Proceedings No. 35, The Chemistry and Chemically Related Properties of Cement*, British Ceramic Soc., 419–429.

Gold, S.J. and Shirvill, A.J. (1992) Effects of metakaolin on concrete strength. Laboratory Report 22 RMC Readymix Limited.

Grube, H. (1985) Influence of concrete materials mix design and construction techniques on permeability. Concrete Society Conference, Permeability of Concrete, London.

Halliwell, M.A. (1992) Preliminary assessment of the performance of Portland cement concrete containing metakaolin. BRE Client Report TCR48/92. BRE.

Hewlett, P. (ed.), *Lea's 'The Chemistry of Cement and Concrete'*, 4th edn. (Chapter 12: 'Microsilica as an addition').

Highley, D.E. (1984) China clay. Mineral Dossier No. 26. HMSO, London.

Hobbs, D.W. (1982) Influence of pulverised-fuel ash and granulated blastfurnace slag upon expansion caused by the alkali-silica reaction. *Concrete Research*, **34**, No. 119, 83–94.

Hobbs, M. (1996) Project Report, University of Surrey.

Holland, T.C. (1983) Abrasion-erosion evaluation of concrete admixtures for stilling basin repairs, Kinzua Dam, Pennsylvania. US Army Engineer Waterways Experiment Station, Structures Laboratory, Vicksburg, Miscellaneous Paper SL-83-16.

Hooton, R.D. (1987) Some aspects of durability with condensed silica fume in pastes, mortars and concretes. International Workshop on Condensed Silica Fume in Concrete, Montreal.

Huang Cheng-Yi and Feldman R.F. (1985) Dependence of frost resistance on the pore structure of mortar containing silica fume. *ACI Journal*, Sept.–Oct.

Hustad. T. and Loland, K.E. (1981) Report 4: Permeability. FCB/SINTEF, Norwegian Institute of Technology, Trondheim, Report STF65 A81031.

Imerys, (2000) Effect of Metastar on the mechanical properties of concrete. Information sheet PMA30/MK.

Johansen, R. (1979) Silicastov i fabrikksbetong. Langtidseffekter. FCB/SINTEF, Norwegian Institute of Technology, Trondheim, Report STF65 F79019.

Johansen, R. (1981) Report 6: long term effects. FCB/SINTEF, Norwegian Institute of Technology, Trondheim, Report STF65 A81031.

Jones, T.R., Walters, G.V. and Kostuch, J.A. (1992) Role of metakaolin in suppressing asr in concrete containing reactive aggregate and exposed to NaCl solution. *Proc. 9th Int. Conf. on AAR*. London.

Jorgen, J.A.H.R. (1980) Possible health hazards from different types of amorphous silicas – suggested threshold limit values. Institute of Occupational Health, Oslo. HD806/79. Revised.

Joshi, R.C. and Lohtia, R.P. (1997) *Advances in Concrete Technology*, Volume 2, *Fly Ash in Concrete*, Gordon and Breach Science Publishers, Now York.

Justesen, C.F. (1981) Performance of dense injection grout for prestressing tendons. Technical Advisory Service, Aalbory Portland, Aalborg, Denmark.

Keck, R.H. and Riggs, E.H. (1997) Specifying fly ash for durable concrete. *Concrete International*, April.

Kostuch, J.A. *et al.* (1993) High performance concretes incorporating metakaolin – a review. *Proc. Int. Conf. Concrete 2000*, Dundee.

Krenchel, H. and Shah, S. (1985) Applications of polypropylene fibres in Scandinavia. *Concrete International; Design and Construction*, 2, No. 3, 32–34.

Kumar, A., Roy, D.M. and Higgins, D.D. (1987) Diffusion through concrete. *Concrete*, **21**, No 1, 31–34, Jan.

Larbi, J.A. and Bijen, J.M. (1991) The role of silica fume and metakaolinite in the Portland cement paste–aggregate interfacial zone in relation to the strength of mortars. PhD thesis, Delft University.

Larbi, J.A. and Bijen, J.M. (1992) Influence of pozzolans on the Portland cement aggregate interface in relation to diffusion of ions and water absorption of concrete. Delft University.

Largent, R. (1978) Research Bulletin. Liaisons LCPC, France.

Larrard, F. *et al.* (1987) (Fracture toughness of high strength concrete. *Proc. Utilisation of High Strength Concrete*, Stavanger, 215–223.

Li, S., Roy, D.M. and Kumar, A. (1985) Quantitative determination of pozzolanas in hydrated systems of cement or Ca(OH)$_2$ with fly ash or silica fume. *Cement and Concrete Research*, **15**, 1079–1086.

Loland, K.E. and Hustad, T. (1981) Report 2: Mechanical properties. FCB/SINTEF, Norwegian Institute of Technology, Trondheim, Report STF65 A81031.

Loland, K.E. (1983) Fasthets-og deformasjonsegenskaper i herdnet tilstand – herdebetingelser. Seminar Bruk av silika i betong, Norsk Sivilingeniorers Forening, Oslo (in Norwegian).

Maage, M. Effect of microsilica on the durability of concrete structures.

Maage, M. and Hammer, T.A. (1985) Modifisert Portlandsement. Delrapport 3. Fasthetsutvikling og E-modul. FCB/SINTEF, Norwegian Institute of Technology, Trondheim, Report STF65 A85041 (in Norwegian).

Maage, M. and Rueslatten, H. (1987) Trykkfasthet og blaeredannelse pa brannpakjent hoyfastbetong. FCB/SINTEF, Norwegian Institute of Technology, Trondheim, Report STF65 A87006 (in Norwegian).

Malhotra, V.M. *et al.* (1987) Mechanical properties and freezing and thawing resistance of high strength concrete incorporating silica fume. International Workshop on Condensed Silica Fume in Concrete, Montreal.

Malhotra, V.M. and Mehta, P.K. (1996) *Advances in Concrete Technology*, Volume 1, *Pozzolanic and Cementitious Materials*, Gordon and Breach Science Publishers, New York.

Markestad, S.A. (1977) An investigation of concrete in regard to permeability problems and factors influencing the results of permeability tests.

Marsh, D. (1992) An alternative to silica fume? *Concrete Products*.

Martin, S.J. (1993a) Suppression of ASR using metakaolin, ggbs, pfa, and microsilica. Laboratory Report 24, RMC Readymix Limited.

Martin, S.J. (1993b) Suppression of ASR using metakaolin. Laboratory Report 33, RMC Readymix Limited.

Martin, S.J. (1995) The use of metakaolin in high strength concrete. Laboratory Report 78, RMC Readymix Limited.

Martin, S.J. (1998) Water demand and other characteristics of metakaolin concrete. Laboratory Report 135, RMC Readymix limited.

Martin, S.J. (1997) Metakaolin and its contribution to the acid resistance of concrete. *Proc. Int. Sym. Concrete for a sustainable agriculture*, Stavanger.

Mass, J.C. (ed.) (1996) Interfacial transition zone in concrete. Rilem Report 11.

Mather, K. (1980) Factors affecting the sulphate resistance of mortars. *Proc. 7th International Conference on Chemistry of Cements*, **4**, 580–585.

Matthews, J.D. and Gutt, W.H. (1978) Studies of fly ash as a cementitious material. Conference on Ash Technology and Marketing, London, October.

McMillan, F.R. and Powers, T.C. (1934) A method of evaluating admixtures, *Proceedings American Concrete Institute*, Vol. 30, 325–344, March–April.

Mehta, P.K. (1981) Sulphate resistance of blended Portland cements containing pozzolans and granulated blastfurnace slag. *Proc. 5th International Symposium on Concrete Technology*, Monterey.

Monk, M.G. (1983), Portland-fly ash cement: A comparison between intergrinding and blending. *Concrete Research*, **35**, 124, September, 131–141.

Monteiro, P.J.M. *et al.* (1985) Microstructure of the steel–cement paste interface in the presence of chloride. *Cement and Concrete Research*, **15**, 781–784.

Monteiro, P.J. *et al.* (1986) Effect of condensed silica fume on the steel–cement paste transition zone. Norwegian Institute of Technology, Trondheim, Report BML 86.205.

Murat, M. (1983) Hydration reaction and hardening of calcined clay and related minerals. Preliminary investigation of metakaolin. *Cement and Concrete Research*, **13**.

Neville, A.M. (1995) *Properties of Concrete* (4th edn), Longman, London.

Neville, A.M. and Brooks, J.J. (1975) Time-dependent behaviour of Cemsave concrete. *Concrete*, **9**, No. 3, 36–39.

Nustone Environmental Trust (2000) The effect of the fineness of PFA (fly ash) on the consistence and strength properties of standard mortar. Project report, March. Read in conjunction with EN450 Fly Ash BS 3892 Part 1 Testing Program, UKQAA, Analysis of results, March 2000.

Okkenhaug, K. (1983) Silikastovets innvirkning pa luftens stabilitet i betong med L-stoff og med L-stoff i kombinasjon med P-stoff. Swedish Cement and Concrete Research Institute at the Institute of Technology, Stockholm CBI report 2:83, 101–105 (in Norwegian).

Okkenhaug, K. and Gjorv, O.E. (1982) Influence of condensed silica fume on the air-void system in concrete. FCB/SINTEF, Norwegian Institute of Technology, Trondheim, Report STF65 A82044.

Osborne, G.J. (1986) Carbonation of blastfurnace slag cement concretes. *Durability of Building Materials*, **4**, 81–96.

Owens, P.L. (1979) Adapted and redrawn from 'Fly ash and its usage in concrete'. *Concrete*, July.

Page, C.L. (1983) Influence of microsilica on compressive strength of concrete made from British cement and aggregates. Elkem Materials Ltd.

Page, C.L. Coleman, N. (1994) Metastar reduced chloride ion penetration. Aston University (data sheet provided by ECCI).

Page, C.L. and Havdahl, J. (1985) Electrochemical monitoring of corrosion of steel in microsilica cement pastes. *Materials and Structures*, **18**, No. 103, 41–47.

Page, C.L. and Vennesland, O. (1983) Pore solution composition and chloride binding capacity of silica fume cement pastes. *Materials and Structures*, **16**, 19–25.

Parker, D.G. (1986a) Alkali aggregate reactivity and condensed silica fume. Ref. No. SA 854/3C. Elkem Materials Ltd.

Parker, D.G. (1986b) Microsilica concrete. Part 2: In use. Concrete Society Current Practice Sheet No. 110, Concrete, March.

Paya, *et al.* (1986) Mechanical treatment of fly ashes, Part II. *Cement and Concrete Research*, **26**, No. 2, 225–235.

Perry, C. and Gillott, J.E. (1985) The feasibility of using silica fume to control concrete expansion due to alkali aggregate reaction. *Durability of Building Materials*, **3**, 133–146.

Pietersen, H.S. *et al.* (1993) Reactivity of flyash and slag in cement. Delft University.

Pigeon and Regourd, M. (1983) Freezing and thawing durability of three cements with various granulated blast furnace slag contents. *Proc. Canment ACI Conference*, ACI publication SP-79, V2, Montebello, Canada, 979–998.

Popovic, K, *et al.* (1984) Improvement of mortar and concrete durability by the use of condensed silica fume. *Durability of Building Materials*, **2**, 171–186.

Ramakrishnan, V. and Srinivasan, V. (1983) Performance characteristics of fibre reinforced condensed silica fume concrete. American Concrete Institute, **11**, 797–812.

Rasheeduzzafar *et al.* Proposal for a code of practice for durability of concrete in the Arabian Gulf environment.

Regourd, M. (1993) *Condensed Silica Fume*. In Aitcin, P.C. (ed.), Université de Sherbrooke, Canada, 20–24.

RMC Readymix Limited (1999) Unpublished data related to the Addashaw Farm project.

Roy, D.M. (1987) Hydration of blended cements containing slag, fly ash or silica fume, Sir Frederick Lea Memorial Lecture, 29 April–1 May 1987, Institute of Concrete Technology annual symposium.

Ryle, R. (1994) Low permeability concrete for the Jubilee Line extension: An assessment of proposed permeability tests, potential mixes and other factors. Confidential Technical Report 257, RMC Readymix Limited.

Saad, M.N.A., de Andrade, W.P. and Paulon, V.A. (1982) Properties of mass concrete containing an active pozzolan made from clay. *Concrete International*, July.

Samuelsson, P. (1982) The influence of silica fume on the risk of efflorescence on concrete surfaces. Norwegian Institute of Technology, Trondheim, Report BML 82.610, 235–244 (in Swedish).

Sandvik, M. (1981) Fasthetsutvikling for silicabetong ved ulike temperaturniva. FCB/SINTEF, Norwegian Institute of Technology, Trondheim, Report STF65 F81016 (in Norwegian).

Sandvik, M. (1983) Silicabetong: herdevarme, egenkapsutvikling. FCB/SINTEF, Norwegian Institute of Technology, Trondheim, Report STF65 A83063 (in Norwegian).

Sellevold, E.J. and Radjy, F.F. (1983) Condensed silica fume (microsilica) in concrete: water demand and strength development. Publication SP-79. American Concrete Institute, **11**, 677–694.

Sellevold, E.J. *et al.* (1982a) Silica fume-cement paste: hydration and pore structure. Report BML 82.610. The Norwegian Institute of Technology, Trondheim, Norway, 19–50.

Sellevold, E.J. *et al.* (1982b) Silica Fume-cement pastes: hydration and pore structure. Norwegian Institute of Technology, Trondheim, Report BML 82.610, 19–50.

Shirley, S.T. *et al.* Fire endurance of high strength concrete slabs. *Concrete International*.

Shirvill, A.J. (1992) Mix design of metakaolin concrete, an interim report. Laboratory Report 23, RMC Readymix Limited.

Sibbick, R.G. and Nixon, P.J. (2000) Paper presented at 11th Int. Conf. on AAR, Quebec.

Singh, B. and Osborne, G.J. (1994) Hydration and durability of OPC/Metakaolin blended concrete. BRE Client Report CR291/94, BRE.

Skurdal, S. (1982) Egenskapsutvikling for silicabetong ved forskjellige herdetemperaturer. Norwegian Institute of Technology, Trondheim, Report BML 82.416 (in Norwegian).

Smith, I.A. (1967) The design of fly ash concretes, ICE, Paper 6982, 769–790.

Smolczyk, H.G. (1977) The use of blastfurnace slag cement in reinforced and prestressed concrete. Sixth International Steel Making Day, Paris.

Sorensen, E.V. (1982) Concrete with condensed silica fume. A preliminary study of strength and permeability. Norwegian Institute of Technology, Trondheim, Report BML 82.610, 189-202 (in Danish).

Swamy, R.N. (ed.) (1986) *Cement Replacement Materials*, Surrey University Press. (Chapter 4 written by Wainwright, P.J.)

Taylor, H.F.W. (1997) *Cement Chemistry* (2nd edn), Thomas Telford Publishing, London.

Thomas, M.D.A (1990) A comparison of the properties of opc and fly ash concrete in 30 year old mass concrete structures. In *Durability of Building Materials and Components*, E & FN Spon, London, 383–394.

Traetteberg, A. and Alstad, R. (1981) Volumstabilititet i blandingssementer med rajernslagg og silikastov. FCB/SINTEF, Norwegian Institute of Technology, Trondheim, Report STF65 A81034 (in Norwegian).

Turrizani, R. (1964) Aspects of the chemistry of pozzolans. In *The Chemistry of Cement*, Chapter 14, Vol. 2, Taylor, London.

Vennesland, O. (1981) Report 3: Corrosion properties. FCB/SINTEF, Norwegian Institute of Technology, Trondheim, Report STF65 A81031.

Vennesland, O. and Gjorv, O.E. (1983) Silica concrete – protection against corrosion of embedded steel. American Concrete Institute, publication SP-79, II, 719–729.

Virtanen, J. (1985) Mineral by-products and freeze–thaw resistance of concrete. Publikation nr. 22:85. Dansk Betonforening, Copenhagen, 231–254.

Wainwright, P.J. and Aider H. (1995) The influence of cement source and slag additions on the bleeding of concrete. *Cement and Concrete Research*, Vol. **25**. No. 7 1445–1456.

Wainwright, P.J. and Tolloczko, J.J.A. (1986) The early and later age properties of temperature cycled opc concretes. *2nd Int. Conf on the use of fly ash silica fume, slag and natural pozzolans in concrete*, CANMET/ACI April 1986, Madrid SP-91 Vol. 2, 1293–1321.

Walters, G.V. and Jones, T.R. (1991) Effect of metakaolin on alkali–silica reactions in concrete manufactured with reactive aggregates. *Proc. 2nd Int. Conf. Durability of Concrete*. Canada.

Wolsiefer, J. (1982) Ultra high strength field placeable concrete in the range 10 000 to 18 000 psi (69 to 124 MPa). American Concrete Institute, Annual Conference, Atlanta.

Wooley, G.R. and Cabrera J.G. (1991) Early age *in-situ* strength development of fly ash concrete in thin shells, Int. Conf. on Blended cement, Sheffield.

Woolley, G.R. and Conlin, R.M. (1989) Pulverised fuel ash in construction of natural draught cooling towers, *Proc. Inst. Civil Engineers*, Part 1 paper 9278, Vol. 86, 59–90.

Xu, H.Y. and Chen, M. (1986) AAR in Chinese engineering practices. *Proc. 7th Int. Conf. on AAR*. Ottawa.

Zhang, M.H. and Malhotra, V.M. (1995) Characteristics of thermally activated alumino-silicate pozzolanic material and its use in concrete. *2nd CANMET/ACI Int. Sym. On Advances in Concrete Technology*.

PART 3

Admixtures

4

Admixtures for concrete, mortar and grout

John Dransfield

4.1 Introduction

4.1.1 Definition and description

Admixtures are chemicals, added to concrete, mortar or grout at the time of mixing, to modify the properties, either in the wet state immediately after mixing or after the mix has hardened. They can be a single chemical or a blend of several chemicals and may be supplied as powders but most are aqueous solutions because in this form they are easier to accurately dispense into, and then disperse through the concrete.

The active chemical is typically 35–40% in liquid admixtures but can be as high as 100% (e.g. shrinkage-reducing admixtures) and as low as 2% (e.g. synthetic air-entraining admixtures). In most cases the added water from the admixture is not sufficient to require a correction for water–cement ratio.

Admixtures are usually defined as being added at less than 5% on the cement in the mix but the majority of admixtures are used at less than 2% and the typical range is 0.3–1.5%. This means the active chemicals are usually present at less than 0.5% on cement or 0.02% on concrete weight.

The dosage may be expressed as litres or kg per 100 kg of cement, and cement normally includes any slag, pfa or other binders added at the mixer.

Admixtures *are not* the same as additives, which are chemicals preblended with the cement or a dry cementitious mix. Neither are they the same as additions, which are added at the mix. Type I additions are essentially inert, e.g. limestone powder or pigments. Type II additions are pozzolanic or latent hydraulic binders such as pfa or silica fume.

4.1.2 Brief history of admixture use

Romans	Retarders	Urine
	Air entrainment	Blood
	Fibres	Straw
Plasticizers	1932	Patent for sulphonated naphthalene formaldehyde plasticizers (but not available in commercial quantities)
	193?	Lignosulphonates used as plasticizers
	193?	Hydroxycarboxcilic acid salts used as plasticizers and retarders
Waterproofers	193?	Fatty acids, stearates and oleates
Air entrainers	1941	Tallow and fatty acid soaps for frost resistance
Superplasticizers	1963	Sulphonated naphthalene formaldehyde commercially available
	1963	Sulphonated melamine formaldehyde patent and available
	1990–1999	Polycarboxylate ether development and introduction

4.1.3 Admixture standards and types

Admixture standards in individual European countries were phased out during 2002 and a new European Standard EN 934 introduced, covering all the main types of admixture. This standard is currently divided into five parts. Part 2, covering concrete admixtures, is probably the most important.

The previous British Standard was BS 5075 Parts 1 to 3 for Concrete and BS 4887 for Mortars. Outside Europe, the American Standard ASTM C494 is widely specified.

See Figure 4.1 for admixture sales by type.

Admixture types covered by EN 934

Normal plasticizing/water reducing (WRA)	EN 934-2
Superplasticizing/high range water reducing (HRWRA)	EN 934-2
Retarding & retarding plasticizing	EN 934-2
Accelerating-set and hardening types	EN 934-2
Air entraining	EN 934-2
Water retaining	EN 934-2
Water resisting (waterproofing)	EN 934-2
Retarded ready-to-use mortar admixtures	EN 934-3
Sprayed concrete	EN 934-5
Grout admixtures for prestressing	EN 934-4

Admixture types not covered by European standards
Corrosion inhibiting
Foamed concrete & low density fill (CLSM)
Polymer dispersions
Pumping aids
Self-compacting concrete
Precast semi-dry concrete
Shrinkage-reducing
Underwater/anti-washout
Washwater

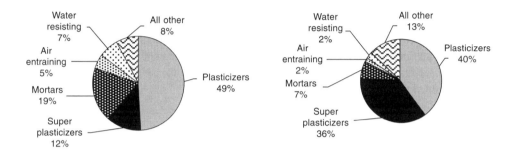

Figure 4.1 UK admixture sales by type (CAA 2000 statistics). These statistics show that normal plasticizers made up the largest proportion of UK admixture sales. This is in contrast to much of Europe where superplasticizers generally take a larger proportion of the market. This difference probably results from the preference of UK buyers to specify 50 mm slump while most other countries specify slumps in excess of 150 mm. Low slumps are generally difficult to place and fully compact and frequently lead to unrecorded addition of water on the job site.

4.1.4 Admixture mechanism of action

Admixtures work by one or more of the following actions:

- Chemical interaction with the cement hydration process, typically causing an acceleration or retardation of the rate of reaction of one or more of the cement phases.
- Adsorption onto cement surfaces, typically causing better particle dispersion (plasticizing or superplasticizing action).
- Affecting the surface tension of the water, typically resulting in increased air entrainment.
- Affecting the rheology of the water, usually resulting in an increased plastic viscosity or mix cohesion.
- Introducing special chemicals into the body of the hardened concrete that can affect specific properties such as corrosion susceptibility of embedded steel or water repellence.

The benefits obtained from using the admixture are often in the hardened properties. However, with the possible exception of the last bullet point, these actions all affect the properties of the wet concrete between the time of mixing and hardening in one or more of the following ways:

- Affect water demand   Plasticizing or water reducing
- Change the stiffening rate   Accelerating/retarding

- Change the air content   Increase (or decrease) entrained air
- Change the plastic viscosity   Cohesion or resistance to bleed and segregation of the mix

One of these effects will usually be the *primary* property, the property for which the admixture is being used. However, the admixture can also affect one or more of the other wet properties. These are called *secondary* effects and it is often these which are key to admixture selection within an admixture type.

For example, a water-reducing admixture may be required to give low water:cement ratio for durability but if the concrete mix lacks cohesion and is prone to bleed, a water-reducing admixture which also increases cohesion would be appropriate to prevent the bleed. The primary function is as a water reducer; the secondary function is to improve cohesion. Incorrect admixture selection could make the bleed worse.

Because of these secondary effects, an admixture which works well on one concrete plant with one source of cement and aggregates could be quite unsuitable on another plant a few miles away where a different aggregate source is being used or a different binder combination is employed.

4.1.5 Rheology and admixtures

The source of the materials used and the mix design determine the basic rheology of a concrete mix. Only after this has been optimized should admixtures be used to modify the rheology.

At the job site, concrete rheology is normally measured by the slump test and concrete technologists may also use their experience to assess the cohesion. Is the mix too sticky or is it likely to segregate or bleed?

In the laboratory, as a measure of the admixture effect, we can use equipment to test for the shear stress against rate of shear strain of a concrete mix. In concrete technology terms the information results in a yield stress which is similar to a slump value and a plastic viscosity which puts a numerical value to cohesion.

From these rheology measurements we can gain an insight into the effect of individual mix constituents as shown in Figure 4.2. Adding water to a mix reduces yield (increases the slump) but also reduces plastic viscosity (cohesion) increasing the risk of bleed and segregation. Dispersing admixtures (plasticizers and superplasticizers) will reduce the

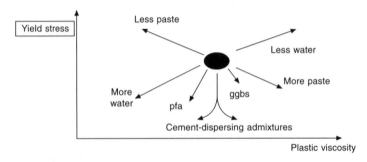

Figure 4.2 Concrete rheology, effect of mix materials (P. Domone).

yield but may increase or decrease the plastic viscosity, depending on the secondary properties of the particular admixture chosen.

It is now possible to look at relative rheology that is required in concrete for various applications and where an admixture is typically used to achieve the requirements (Figure 4.3).

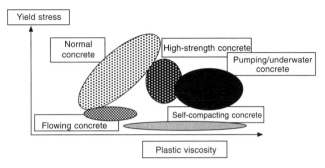

Figure 4.3 Concrete rheology by application (P. Domone).

High-strength concrete is often cohesive (high plastic viscosity) because of the high cement and low water content. Underwater and pump concrete also need to be cohesive to resist segregation and admixtures are often used to achieve this. As the yield stress falls, the plastic viscosity of normal concrete also falls unless the mix design is modified to increase cohesion and prevent segregation, as required by high-slump, flowing concrete. In self-compacting concrete, yield stress is almost zero but a range of plastic viscosities is required to meet applications from a rapid flow, wet mix to a slow creeping flow with high plastic viscosity but still having a low yield stress.

Using this information the effects of individual admixtures or admixture combinations can be selected to achieve the required changes in yield value and plastic viscosity for a particular job function.

4.1.6 Cement and concrete chemistry in relation to admixtures

Proportions of concrete mix constituents:	*By weight*	*By volume*
• Cement	15%	11%
• Water required for cement hydration	3%	8%
• Additional water to give workability	5%	11%
• Aggregate	77%	70%

The most important factor in this typical mix design is the 11% by volume of water that is added, only to give the mix workability. This water does not react with the cement but remains unbound and free in the concrete after hardening. This forms interconnecting water-filled voids called capillaries that reduce strength, increase permeability to water and provide paths for diffusion of chlorides, sulphates and other aggressive chemicals. As the water slowly evaporates from the capillaries, it causes drying, shrinkage and cracking.

The empty capillaries provide a rapid path for carbonation.

Water-reducing admixtures can significantly reduce the volume of this unbound water leading to stronger and more durable concrete.

Size of concrete mix constituents are typically as follows:

- Aggregate   20.0–4.0 mm
- Sand   4.0–0.1 mm
- Entrained air   0.3–0.05 mm
- Cement   0.075–0.01mm
- Capillaries   < 0.001 mm
- Gel pores   <0.00003 mm
- Admixture   < 0.00001 mm

Entrained air voids are typically four times the size of a cement grain and are similar in size to the lower third of a typical sand grading. Air entrainment/stability and sand grading/content are closely linked.

Capillaries are less than one tenth the diameter of a cement grain but account for 10% of the concrete volume – there must be a lot of them! Admixtures are less than a thousandth the diameter of a cement grain. They adsorb onto active sites on the grain.

Cement hydration starts as soon as water is added to the mix. Water absorbs into the outer part of the cement grain, dissolving calcium and hydroxide ions, which move out into the surrounding water. This leaves the cement surface with a calcium-depleted hydrosilicate layer carrying a negative charge. Some of the positive calcium ions in solution then adsorb back onto this silicate surface to give a positive charge as measured by the Zeta potential. Water continues to diffuse into the grain, releasing further calcium and hydroxide ions into solution and increasing the thickness of the hydrosilicate layer.

This is the predominant reaction between the cement and the water and mainly involves the tricalcium silicate (C_3S) phase of the cement. However, other phases are also present, including tricalcium aluminate (C_3A), which is very reactive although present at only about 10% of the cement weight. Sulphate, in the form of gypsum, is added to the cement to help control the C_3A reactivity by forming ettringite. Ettringite has a reactive surface that can attract and adsorb a disproportionate amount of some admixtures. For this reason, low C_3A (sulphate-resisting or ASTM Type V cements) may require a slightly lower dosage of admixture for equivalent effect.

Some admixtures can affect the rate of sulphate solubility, especially if the gypsum becomes dehydrated to hemi-hydrate during cement grinding. This and some other changes in cement chemistry, which do not affect the cement properties on their own, can result in abnormal setting in the presence of admixtures. This abnormal cement admixture interaction is very rare but embarrassing when it happens. It usually occurs soon after a delivery of very fresh cement.

When water is added to cement in a normal concrete mixer, the cement grains are not uniformly dispersed throughout the water but tend to form into small lumps or flocs. These flocs trap water within them and so the mix is less mobile and fluid than would be the case if the cement were in the form of individual grains. High-shear grout mixers can break up these flocs producing a more fluid mix but there is a tendency for them to re-aggregate. Dispersing admixtures are used to both break up the flocs and to maintain the dispersion.

4.2 Dispersing admixtures

Dispersing admixtures are the most important admixture type, accounting for over 60% of all admixtures sold in the UK. They include the types covered by the names Plasticizer, Superplasticizer, Water Reducer, High Range Water Reducer and they are covered by Standards EN 934-2 or by ASTM C494. Marketing ploys by admixture companies have introduced additional/alternative names including Mid-range, first, second and third generation, hyperplasticizer, etc. However, these names are generally not recognized by standards.

These dispersing admixtures all function by adsorption onto the cement surface in a way that causes the cement particles to distribute more uniformly throughout the aqueous phase, reducing the yield value for a given water content and so increasing the fluidity of the mix. They may also have a small effect on the plastic viscosity or cohesion of the mix. They are used in one or more of the following ways:

* To increase the fluidity of the mix at a given water content. (Plasticizing or superplasticizing action)
* To reduce the water content of the mix at a given fluidity, improving strength and durability. (Water reducing or High Range Water Reducing)
* To reduce the cement content by reducing water and maintaining water:cement ratio to give equivalent strength. (Cement reducing)

In optimizing a concrete mix for a given application it will often be effective to use a combination of these properties (for example, to use the admixture at a dosage which can reduce the water content but also allow the fluidity of the mix to increase).

Dispersing admixtures fall into to two groups which, for standards purposes (EN 934 or ASTM C494), are divided by their water-reducing ability. These are the Normal Plasticizers/Water reducers and the Superplasticizers/High Range Water reducers.

4.2.1 Normal plasticizers

Definition: water reduction more than 5% but less than 12%.
Other characteristics:

* Low dosage, typically 0.2–0.6% on cement (30–40% soln).
* Often have significant secondary effects in the upper half of the dosage range, especially retardation with some binder blends and/or at lower temperatures.
* Higher dosages give little additional dispersion but very significant secondary effects.

Plasticizers are usually based on lignosulphonate, which is a natural polymer, derived from wood processing in the paper pulp industry. The brown lignin in the wood pulp is removed by a sulphite reaction and can then be processed to remove sugars and some other unwanted material before being used for admixtures. The greater the level of processing, the fewer the secondary effects.

Other dispersants may be blended with the lignin, e.g. gluconate, maltodextrin etc. These can boost certain properties but also tend to increase the potential for retardation.

The admixture may also be modified with air entrainers, air detrainers and small amounts of other chemicals to boost or control secondary effects.

A more recent development is normal plasticizers based polycarboxylate ether. This chemical was originally developed as a superplasticizer, but supplied as a dilute solution to decrease sensitivity; it can also be effective as a normal or multi range plasticizer and gives the advantage over lignosulphonates of having minimal secondary effects.

Normal plasticizers are extensively used by readymix companies to optimize their mix designs, especially at low to medium slump (up to about 130 mm). They account for almost 50% of all admixture sales in the UK but a lower proportion in most other countries where initial slump values tend to be significantly higher.

4.2.2 Superplasticizers

Definition: water reduction more than 12% but depending on dose and type can give over 30%.

Other characteristics:

- Dosage, typically 0.6 to 2.0% on cement (30–40% soln).
- Minimal retardation except at the top of the dosage range.
- Synthetic chemicals derived from the chemical industry and designed with properties optimized specifically for admixture applications. There are currently three main chemical types: the sulphonated polymers of naphthalene or melamine formaldehyde condensates and the polycarboxylate ethers. These will be discussed in more detail later in this section.

  These basic chemicals can be used alone or blended with each other or with lignosulphonate to give admixtures with a wide range of rheological properties. They can also be blended with other chemicals to increase retardation, workability retention, air entrainment and other properties.

Superplasticizers are a versatile group of admixtures with a wide range of potential properties and dosages, tailored to meet specific requirements. Dual-function products that are designed to give, say, water reduction and retardation are common, especially in warmer climates. It is therefore essential to check with the data sheet and the manufacturer before moving from one type or brand of superplasticizer to another.

Superplasticizers find use in the pre-cast and readymix concrete industries as well as in site-mixed concrete. In pre-cast they are used mainly as water reducers to give high early strength. On site-mixed they may be used:

- To give high workability where there is dense reinforcement
- For low water contents to give high early or later age strength
- For low w:c to give low-permeability durable concrete.

In readymix they are used for all the above applications and can also give extended workability to cope with long delivery and placing times, especially at elevated temperatures.

4.2.3 Chemical structure of dispersing admixtures

Lignin is a natural polymer found in wood and is of variable structure, depending on the timber source and time of year when the wood is cut. In one paper pulping process, the brown lignin is sulphonated to make it soluble prior to extraction as a by-product. To make this lignosulphonate suitable for use as a cement dispersant, it is then chemically modified by hydrolysis or fermentation to remove some of the carbohydrate material. The key part of the lignosulphonate can be regarded as sulphonated phenyl propane units of which a very simplified structure is shown in Figure 4.4. Potential ether linkages to further phenyl propane units to make up the polymer structure are indicated.

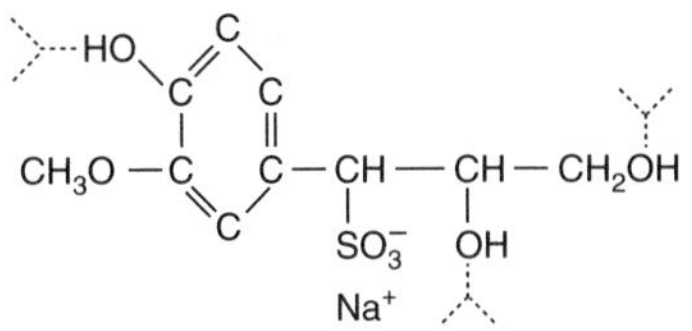

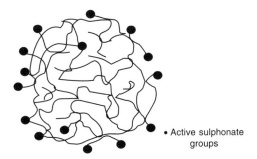

Figure 4.4 Simplified, sulphonated phenyl propane units showing some potential polymer crosslinking points.

The polymeric lignosulphonate structure, represented in Figure 4.5, is a complexed and heavily crosslinked spherical microgel with a broad molecular weight distribution ranging from about 2000 to 60 000. In solution the active sulphonate groups dissociate to $-SO_3^-$ and Na^+ leaving the microgel with an overall negative charge which is key to its dispersing properties.

• Active sulphonate groups

Figure 4.5 Lignosulphonate microgel, less than one-thousandth the diameter of a cement particle.

Sulphonated naphthalene formaldehyde condensate (SNF) also called polynaphthalene sulphonate (PNS) is derived from the chemical industry. Petroleum or coal tar naphthalene is sulphonated using very concentrated sulphuric acid at high temperature and is then polymerized with formaldehyde and neutralized to the sodium (Na) or calcium (Ca) salt. The polymers are of relatively low molecular weight, the sulphonated naphthalene repeating unit **n** shown in Figure 4.6 typically being in the range 2–10 to give molecular weights in the order 500–2500. The higher molecular weights generally give better properties.

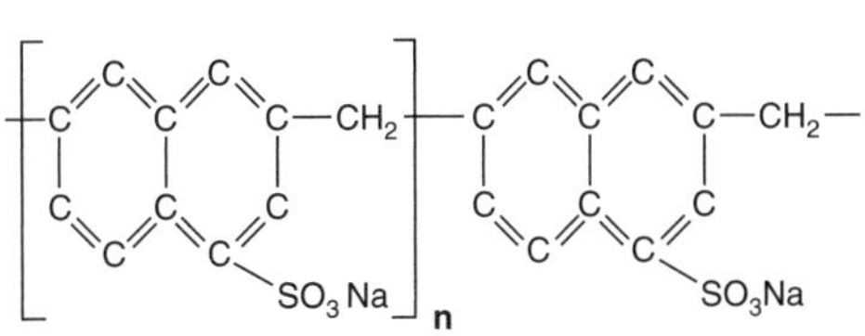

Figure 4.6 Sulphonated naphthalene formaldehyde condensate polymer.

The chain is linear and is free to rotate about the formaldehyde derived CH_2 group so the $-SO_3Na$ can be above or below the chain axis. In solution, like lignosulphonate, the

$-SO_3Na$ dissociates to $-SO_3^-$ and Na^+. The negative charges on the $-SO_3^-$ are the key to the admixture adsorbing onto the cement and also to its electrostatic dispersion mechanism.

Sulphonated melamine formaldehyde condensates (SMF) are very similar in structure except that a melamine ring replaces the naphthalene double ring and the molecular weight is higher (Figure 4.7). They are only available as the sodium salt.

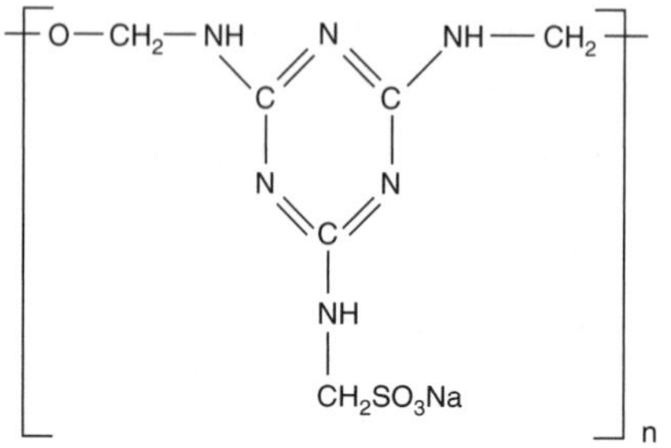

Figure 4.7 Sulphonated melamine formaldehyde condensate polymer.

Polycarboxylate ethers (PCE) are also called PCs or comb polymers. They are the most recent development and unlike SNFs and SMFs, which are essentially a single structure, PCEs are a family of products with significantly different chemical structures. A typical PC structure is shown in Figure 4.8.

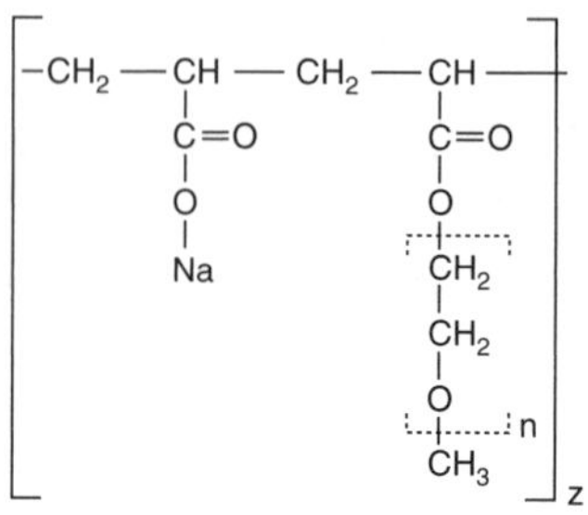

Figure 4.8 Basic structure of a polycarboxylate ether superplasticizer.

The backbone polymer is typically based on polymerization of acrylic acid but this can be substituted or replaced with other monomer groups and can be used to modify the number of carboxylate groups along the polymer backbone. The carboxylate group is normally neutralized as the sodium salt and takes on a negative charge in solution as the Na^+ dissociates. This provides the attachment point for the admixture to adsorb onto the cement surface.

The co-polymer is a polyether shown here as polyethylene glycol ($-CH_2-CH_2-O-$)n. Other polyethers or combinations of polyether can be used and the molecular weight varied, n typically ranging from 20 to 80 units. This together with the number of polyether groups substituted along the chain and the length of the chain can give a significant range of properties. This allows the basic copolymer to be tailored for, say, high early strength for pre-cast or for workability retention for readymix even before other chemicals are blended in. The polyether is the part responsible for the dispersion of the cement particles and works by a steric effect.

4.2.4 Dispersing mechanism

When water is added to cement, the grains are not uniformly dispersed throughout the water but tend to form into small lumps or flocs. These flocs trap water within them causing the mix to be less mobile and fluid than would be the case if the cement were in the form of individual grains. This is analogous to the situation where people are walking both ways along a narrow pavement. If groups of people walking in one direction hold hands, then it is hard for people walking the other way to pass, especially if they are also holding hands. If everyone stops holding hands and walking in groups, then it is easy for people going in opposite directions to move round each other and pass.

Admixtures adsorb onto the cement surfaces and break up the flocs, leaving individual cement grains, which can pass each other easily, making the mix more fluid (Figure 4.9).

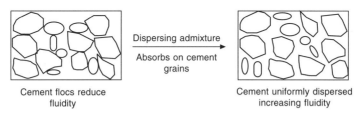

Dispersing admixture
Absorbs on cement grains

Cement flocs reduce fluidity

Cement uniformly dispersed increasing fluidity

Figure 4.9 Effect of dispersing admixtures in breaking up cement flocs.

There is no change in volume of the mix, just the release of water trapped within the floc and the ability of individual cement grains to reorientate to make more efficient use of all the available volume without inter-particle contact. Not only has the fluidity increased, but with water surrounding the whole grain, hydration is more efficient.

If water is now removed from the system, the fluidity can be reduced back to what it was before the admixture addition. However, the average inter-grain distance will also reduce, so less hydration will be necessary before the space between the cement grains is filled with hydration product to give setting and strength development. This leads first to a higher early strength. With continued hydration, more inter-grain space eventually becomes filled with hydration products so late age strength is also improved. Finally, there will be less void space, not filled with any hydration product resulting in fewer capillaries and therefore better durability.

Admixture adsorption onto the cement surface is through the negative charges on the admixture onto the positively charged calcium ions on the cement surface. The dispersion is then by one of two mechanisms, electrostatic repulsion or steric stabilization.

Electrostatic dispersion is the main mechanism in lignosulphonates and in SNF/SMF superplasticizers. These molecules all carry $-SO_3Na$ groups, which in water dissociate into $-SO_3^-$ and Na^+. The $-SO_3^-$ remains attached to the admixture and carries a strong negative charge. Part of this charge is used to attach the admixture to the cement but the remainder orientates out from the grain and repels the negative charges on admixture adsorbed onto adjacent cement grains, causing them to move and stay apart. See Figure 4.10.

As well as seeing the increased fluidity, the effect can be followed by looking at Zeta potential measurements. These measure the charge on the cement surface, which is positive before the admixture addition but then goes strongly negative as the admixture adsorbs.

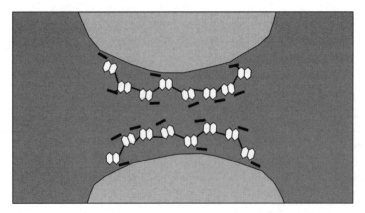

Figure 4.10 Electrostatic dispersion of cement grains by a SNF superplasticizing admixture. The real admixture molecule is much smaller, relative to the cement grain, than is shown.

(See Figure 4.11). The magnitude of the change in the charge is much greater with the superplasticizers than with normal plasticizers, helping to explain the difference in performance.

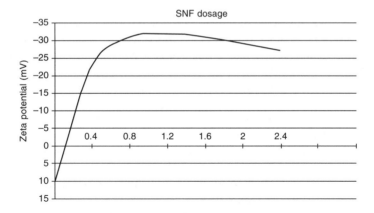

Figure 4.11 Typical effect of zeta potential (surface charge on cement grain) with addition of an SNF superplasticizer.

Steric stabilization is the main mechanism in PCE superplasticizers. These molecules carry $-CO_2Na$ groups which, in water, dissociate into $-CO_2^-$ and Na^+. The $-CO_2^-$ remains attached to the admixture and carries a moderate negative charge that is used to attach the admixture to the cement. The long polyether groups orientate away from the cement surface but will resist becoming entangled with the polyether chains attached to an adjoining cement grain, thus keeping the two grains apart. Zeta potential measurements show that, unlike SNF, there is no significant negative charge on PCE superplasticized cement (Figure 4.12).

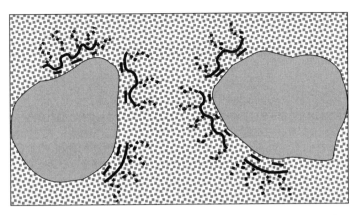

Figure 4.12 Steric stabilization dispersion of cement grains by a PCE superplasticizing admixture. The real admixture molecule is much smaller, relative to the cement grain, than is shown.

4.2.5 Normal plasticizer performance and applications

Admixture performance always depends on the concrete materials and mix design being used as well as the dosage of the admixture. Normal plasticizers may be modified to enhance performance when slag or pfa binders are being used; they may also contain an air detrainer to control the air content of the mix. These and other modifications can affect cohesion, retardation, etc. and should be taken into account during admixture selection.

Typically, normal plasticizers will give 8–10% water reduction. Depending on retardation this will translate to a 120% strength improvement at 24/36 hours and 112% at 28 days. It is difficult to improve on this water reduction but with a slightly higher dose, a higher workability can be additionally obtained.

Most normal plasticizers will increase the air content by between 1% and 2%, which will give some additional cohesion to the mix and may reduce any tendency to bleed especially if the plasticizer is being used to increase workability. The higher the air content of the mix without admixture, the greater will be the additional air that is entrained by an admixture.

Most normal plasticizers give some retardation, 30–90 minutes over a control mix. However, this can become more significant with products that have been modified with molasses, maltodextrin, hydroxylated polymers and other similar chemicals. The effect may only become pronounced at higher dosages or lower temperatures and or where higher levels of a slag binder have been used. It can be a particular problem in slabs, where the low dosage of these admixtures contributes to non-uniform mixing and results in patches of concrete with different levels of retardation that are difficult to finish with power-trowelling equipment and can lead to delamination in the hardened floor.

The main application of normal plasticizers is in readymix to optimize (reduce) the cement content by water reduction at constant w:c. They are also effective in providing increased initial workability which, assisted by any increase in retardation, helps with workability retention where long delivery times may occur.

Compared to a control mix without admixture:

- A normal plasticizer used for water reduction will give some improvement in other properties including permeability, chloride diffusion, shrinkage, etc.

- Used for cement reduction at constant/slightly reduced w:c the effects are more or less neutral.
- Used only to increase workability, most properties including strength will be slightly reduced.

4.2.6 Superplasticizer performance and applications

Superplasticizers can be supplied as pure products, SNF, SMF or PC, or they can be blended with each other or with lignosulphonate as well as with other modifying chemicals. In addition the concentration may vary to a much greater extent, 15–40%, than is usually found in normal plasticizers which are more typically 30–42%. This means that the range of properties that can be obtained is much greater and the following notes are only of general guidance: there will be exceptions with some products.

Sulphonated melamine formaldehyde condensates (SMF) give 16–25%+ water reduction. They tend to reduce cohesion in the mix, increasing the tendency to bleed and segregate but can be very effective in improving mixes which, because of the materials/mix design, are already over-cohesive and sticky. SMF gives little or no retardation, which makes them very effective at low temperatures or where early strength is most critical. However, at higher temperatures, they lose workability relatively quickly and are often added at the site rather than during batching at a readymix plant. SMF generally give a good finish and are colourless, giving no staining in white concrete. They are therefore often used where appearance is important.

Sulphonated naphthalene formaldehyde condensates (SNF) typically give 16–25%+ water reduction. They tend to increase the entrapment of larger, unstable air bubbles. This can improve cohesion but may lead to more surface defects. Retardation is more than with SMF but will still not normally exceed 90 minutes over a control mix, even at the highest dosages. Workability retention is quite good but can be significantly improved if modified with other plasticizers/retarders. SNF is a very cost-effective and forgiving superplasticizer and has found widespread use in all parts of the industry especially in readymix and site-mix applications.

Polycarboxylate ether superplasticizers (PCE) typically give 20–35%+ water reduction. They are relatively expensive per litre but are very powerful so a lower dose (or more dilute solution) is normally used. As previously noted, the basic molecule can be manipulated to give targeted properties. This has currently led to products targeted at pre-cast, having very good water reduction, little retardation but only moderate workability retention. Other formulations target readymix with enhanced workability retention but more normal levels of water reduction and retardation. The PCEs are the most recent superplasticizer development and are still establishing their position in the market, with development continuing to optimize their performance. They have already gained a dominant position in the pre-cast market and for self-compacting concrete but can be more expensive and less forgiving of mix design and material consistency than SNF for general readymix and site use, but even here the PCE market share is growing.

The main benefits of superplasticizers can be summarized as follows:

- Increased fluidity for: Flowing, self-levelling, self-compacting concrete
                              Penetration and compaction round dense reinforcement

- Reduced w:c for: Very high early strength, >200% at 24 hours or earlier
Very high later age strengths, >100 MPa.
Reduced shrinkage, especially if combined with reduced cement content.
Improved durability by removing water to reduce permeability and diffusion

Superplasticizers are not as cost effective as cement reducers.

4.3 Retarding and retarding plasticizing/superplasticizing admixtures

Definition: extend the time for the mix to change from the plastic to the hardened state by at least 90 minutes but not more than 360 minutes at the compliance dosage.

If also plasticizing/water reducing then should also give more than 5% but normally less than 12% water reduction. If also superplasticizing/high range water reducing then should also give more than 12% water reduction but, depending on dose and type, can give significantly more.

Dosage is typically 0.2–0.6% on cement if only supplied as a retarder. If it is a multifunction admixture with dispersing properties, the dosage will be in the same range as for plasticizing/superplasticizing admixtures.

Most retarding admixtures are sold as multifunction retarding plasticizing/superplasticizing products. They are based on the same chemicals as the dispersing admixtures but have molasses, sugar or gluconate added to provide the retardation. Pure retarders are usually based on gluconate and or sucrose although these also plasticize. Phosphates are occasionally used and retard with little plasticizing action.

Significant overdosing of most retarding admixtures quickly lead to large extensions of set time running into days and in extreme cases can result in the cement never properly hydrating and gaining strength. Phosphates are more tolerant to overdosing but are less effective in most applications and are rarely used.

Temperature has a significant effect on retardation and this needs to be taken into account when selecting the dosage or running trial mixes. Even a 5° temperature change can have a large effect. Retardation also increases if pfa or ggbs form part of the cement content and should also be taken into account, especially if the temperature is likely to be low.

Retarding admixtures find use in readymix and on-site concrete to help provide workability retention and/or to prevent cold joints on large or delayed pours.

4.3.1 Mechanism of retardation

The mechanism of setting of cement is not fully understood and this impacts on our understanding of the way retarding admixtures work. There are probably two main processes involved:

- A blocking mechanism where the admixture adsorbs strongly on the cement surface, slowing the formation of silicate hydrates.

- Chelation of calcium ions in solution, preventing the precipitation of calcium hydroxide (portlandite).

One or both of these processes may be involved, depending on the admixture type selected. The latter is probably the more important for most retarder types and prevents setting but not workability loss. This is important to appreciate if the required performance is to be achieved from a retarding admixture.

4.3.2 Workability retention

Cement hydrates by progressive absorption of water into the grain followed by the formation of calcium silicate hydrates (CSH). The total water required for hydration is typically in the order of 70–90 litres per cubic metre of concrete. While some of this water will absorb during the initial mixing, further absorption occurs during the plastic stage. Each 20 litres of water lost into the cement grain will roughly half the workability. Combine this with the increased interaction between the hydration products being formed and it is apparent why there is progressive workability loss.

Retarding admixtures do not slow the rate of water absorption and do not prevent all the hydration reactions so they have little effect on workability loss.

The best way to have the required workability at a given time after mixing is to use a plasticizer/superplasticizer to increase the initial workability to a point where the workability loss will give the required slump at the required time (Figure 4.13).

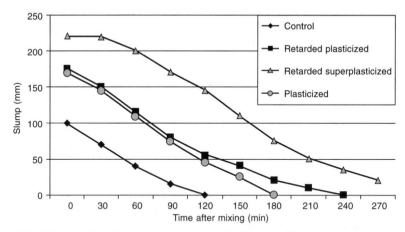

Figure 4.13 Initial workability of concrete and typical workability loss with time.

If the high workability is required close to or after the time of initial set of the concrete, then a retarder will be required in addition to the plasticizer in order to slow down the hydration and setting reactions of the cement. However, as many plasticizers also slightly retard as a secondary effect, there may be no requirement for the retarding admixture. In Figure 4.13 if a slump of 100 mm were required at 90 minutes then a normal plasticizer would probably meet the requirements.

4.3.3 Set retardation

When water is first added to cement there is a rapid initial hydration reaction, after which there is little formation of further hydrates for typically 2–3 hours. The exact time depends mainly on the cement type and the temperature. This is called the dormant period when the concrete is plastic and can be placed. At the end of the dormant period, the hydration rate increases and a lot of calcium silicate hydrate and calcium hydroxide is formed relatively quickly. This corresponds to the setting time of the concrete.

Retarding admixtures delay the end of the dormant period and the start of setting and hardening. This is useful when used with plasticizers to give workability retention. Used on their own, retarders allow later vibration of the concrete to prevent the formation of cold joints between layers of concrete placed with a significant delay between them.

4.3.4 Retarder performance and applications

The use of retarding admixtures to assist plasticizers/superplasticizing admixtures in giving workability retention over extended periods beyond the normal setting time of the mix is well covered above. Workability retention is required in order to place and fully compact the concrete.

Cold joints are formed when fresh concrete is placed onto a previous pour that has already undergone an initial set, preventing the two pours being vibrated into a single monolith. It typically occurs on large or complicated pours when, after placing one section of concrete, adjacent areas of concrete are placed over an extended period before further concrete is placed onto or adjacent to the first pour. Over this time period, the first concrete pour has not only lost workability but has started to set so that it is no longer affected by the action of a vibrator. The new batch of concrete fails to fully bond to the earlier pour and a line of weakness with respect to both strength and permeability now exists in the structure. The first batch of concrete does not need workability for the two pours to be vibrated but it must not have gone beyond its initial set. A retarder can be used to achieve this, and in extreme cases can delay the initial set for over 24 hours for very difficult or large pours.

Cold joints often occur as a result of delays and breakdowns on-site and this possibility should be taken into account when planning any large job. Readymix supply can be interrupted by a traffic accident or by a plant breakdown, on-site a pump blockage or breakdown can also cause a serious delay in the placing of concrete. If this results in a cold joint in a bridge deck or a slip form, the resulting problems can be extremely embarrassing.

Mass concrete pours can also cause problems with cold joints. Unretarded concrete, placed in the early stages of the pour, will start to set and exotherm after a few hours. The heat liberated will accelerate the set of the concrete above and can eventually affect the active surface, resulting in an unexpectedly quick set and formation of a cold joint. The use of a retarder in the early pours with a progressively reduced dosage as the pour progresses can solve this problem.

Retarders do not reduce the exotherm or the peak temperature of mass concrete. They only delay the onset of the exotherm. The only effective way to control exotherm is by the use of less reactive binders such as pfa or ggbs. Chilled mixing water or cooling the aggregate are also effective ways of reducing the peak temperature.

Accidental overdosing of a retarding admixture can significantly delay the setting and strength development. As soon as the error is spotted, the following actions should be taken:

- Protect the surface from evaporation by effective curing
- Revibrate the concrete up to the time of initial set to prevent settlement cracking
- Contact the admixture manufacturer for advice on the likely retardation time at the actual dosage used.

Unless an extreme overdose was used, most concrete will set and gain strength normally within 3–5 days. Beyond 5 days, the potential for the concrete not to gain full strength increases significantly.

4.4 Accelerating admixtures

Accelerating admixtures can be divided into groups based on their performance and application:

- Set accelerating admixtures, reduce the time for the mix to change from the plastic to the hardened state. They can be subdivided into two groups:
  - Sprayed concrete accelerators, which give very rapid set acceleration (less than 10 minutes) and will be covered in section 4.11.
  - Concrete set accelerators which, according to EN 934-2, reduce the time for the mix to change from the plastic to the hardened state by at least 30 minutes at 20°C and at least 40% at 5°C.
- Hardening accelerators, which increase the strength at 24 hours by at least 120% at 20°C and at 5°C by at least 130% at 48 hours.

Other characteristics are:

- Dosage, typically 0.5–2.0% on cement.
- Most accelerators are inorganic chemicals and in most cases are either set or hardening accelerators but not both, although blends are available which provide a degree of both properties.

Calcium chloride is the most effective accelerator and gives both set and hardening characteristics. Unfortunately, it also reduces passivation of any embedded steel, leading to potential corrosion problems. For this reason, it should not be used in concrete where any steel will be embedded but may be used in plain unreinforced concrete.

Chloride-free accelerators are typically based on salts of nitrate, nitrite, formate and thiocyanate. Hardening accelerators are often based on high range water reducers, sometimes blended with one of these salts.

Accelerating admixtures have a relatively limited effect and are usually only cost-effective in specific cases where very early strength is needed for, say, access reasons. They find most use at low temperatures where concrete strength gain may be very slow so that the relative benefit of the admixture becomes more apparent.

4.4.1 Mechanism of acceleration

Like retarders, the mechanisms are not well understood. Set accelerators appear to accelerate the formation of ettringite. Inorganic hardening accelerators increase the rate of dissolution of the tricalcium silicate, leading to an increase in calcium silicate hydrate (CSH) at early ages. Hardening accelerators based on high range water reducers reduce the distance between cement grains so that, for a given amount of CSH hydration product, there is more interaction between the cement grains and hence more strength.

4.4.2 Accelerator performance and applications

Set accelerators have relatively limited use, mainly to produce an early set on floors that are to be finished with power tools. They can, for instance, be used only in the later mixes so that these are accelerated and reach initial set at the same time as earlier pours, allowing the whole floor to be finished at the same time.

Hardening accelerators find use where early stripping of shuttering or very early access to pavements is required. They are often used in combination with a high range water reducer, especially in cold conditions.

At normal ambient temperatures, hardening accelerating admixtures are rarely cost-effective and a high range water reducer is usually a more appropriate choice. The relative effects are shown in Table 4.1.

Table 4.1 Comparison between accelerating admixtures and a high range water reducer

Temperature	5°C Accelerator	20°C	
		Accelerator	High range water reducer
Final set, hours	5.5	2.25	4.75
6 hours, strength N/mm^2	0.0	7.0	2.5
8 hours	1.8	12.5	8.5
12 hours	3.5	20.0	23.5
24 hours	12.5	27.5	35.0
28 days, strength N/mm^2	48.5	50.5	63.0

In summary, a hardening accelerator may be appropriate for strength gain up to 24 hours at low temperature and up to 12 hours at ambient temperatures. Beyond these times, a high range water reducer alone will usually be more cost-effective.

4.5 Air-entraining admixtures

These are defined in EN 934-2 as admixtures that allow a controlled quantity of small, uniformly distributed air bubbles to be incorporated during mixing and which remain after hardening. Air entrainment is usually specified by the percentage of air in the mix but may also be specified by air void characteristics of spacing factor or specific surface of the air in the hardened concrete.

Dosage is typically 0.2–0.4% on cement weight unless they are supplied as a multifunctional product, usually with a dispersing admixture. Air-entraining admixtures are very powerful and may only contain 1–5% solids. The low dose and powerful action require good, accurate and well-calibrated dosing equipment if a consistent level of air is to be achieved.

Strength is reduced by entrained air, typically 5–6% reduction for each 1% of additional air. For this reason, air should normally be limited to the lowest level necessary to achieve the required properties. However, in low cement content, harsh mixes, air may improve workability allowing some water reduction that can offset part of the strength loss.

Air entrainment is used to produce a number of effects in both the plastic and the hardened concrete. These properties will be considered in more detail later in the section but include:

- Resistance to freeze–thaw action in the hardened concrete
- Increased cohesion, reducing the tendency to bleed and segregation in the plastic concrete
- Compaction of low workability mixes including semi-dry concrete
- Stability of extruded concrete
- Cohesion and handling properties in bedding mortars

Air entrainment is very sensitive to the concrete mix constituents, mixing, temperature, transport, pumping etc. and achieving consistent air content from batch to batch can be difficult if these parameters are not well controlled.

Air-entraining admixtures are surfactants that change the surface tension of the water. They have a charged hydrophilic end that prefers to be in water and a hydrophobic tail, which traps air bubbles as a way of getting out of the water. Traditionally, they were based on fatty acid salts or vinsol resin but these have largely been replaced by synthetic surfactants or blends of surfactants to give improved stability and void characteristics to the entrained air.

4.5.1 Factors affecting air entrainment

The sand content and grading undoubtedly has the greatest influence on air entrainment. It is almost impossible to entrain a significant amount of air into a neat cement paste (except by the use of a pre-foam system). As the sand content increases, so does the amount of air under otherwise equal conditions, with an optimum level being reached at about 1:3 cement to sand. Entrained air is very similar in size to the bottom third of most sand gradings (see relative size table in section 4.1.6) and is probably stabilized by slotting into appropriate sized holes in the grading curve.

Sand grading also has a significant effect as can be seen in a control mix with no admixture. For a given mix design, different sand sources with different gradings can change the air content from about 0.6% to over 3%. The more a sand tends to entrain air, the higher the air content that will be entrained by an admixture at a given dose (note that this also applies to plasticizers that may have air entrainment as a secondary effect).

Cement fineness and dust in aggregates also affect stability, probably by clogging up gaps between sand grains, reducing suitable voids for bubble formation and by absorption of free water from the bubble surfaces.

In relation to production of consistent air-entrained concrete, changes in cement content and fineness should be avoided and sand grading should also be consistent. Dust in any of the aggregates should be avoided. Because air is similar in size to the finer sand, it acts in a similar way, increasing cohesion. In already cohesive mixes (high cement content or fine sand) it may be necessary to reduce the sand content of the mix to reduce cohesion but if the cement:sand ratio falls to about 1:2 problems with consistency and stability of air content may be encountered.

The use of pfa in air-entrained concrete can give significant problems in achieving the required air content and batch-to-batch consistency, often resulting in low air contents and/or much higher than expected admixture dose. This is due to the presence of partially burnt coal, which has an active surface that adsorbs the surfactants used as air-entraining admixtures. The 'Loss on Ignition' figure for the pfa may not be a guide to potential problems as this test also picks up unburned coal, which does not adsorb the surfactant. Some types of air-entraining admixture are more tolerant of pfa than others but none are immune to the problem, so great care and regular checking is essential, especially when taking delivery of a fresh batch of pfa.

Entrained air is reasonably tolerant of continued agitation but can be lost if subjected to delivery in a non-agitated vehicle such as a tipping lorry. Air can also be lost during pumping. It is therefore best to check the air content at the final point of discharge into the shutter and adjust the admixture to achieve the required amount of air at that point. Finally, over-vibration can also reduce the air content and may additionally result in a foam of air on the top surface. It should be noted that the higher the workability of the concrete, the greater is likely to be the loss of air from any of these factors.

4.5.2 Freeze–thaw resistance

Frost damage to concrete results from water within the capillary system freezing and expanding to cause tensile forces which crack the concrete or cause surface scaling. It should be remembered that the capillaries can amount to about 10% of the concrete volume and are the result of the excess water added to the concrete to give workability, see section 4.1.6. For frost damage to occur, the concrete must be almost saturated (capillaries full of water).

In air-entrained concrete, air voids are formed that intersect the capillaries at regular intervals, preferably at no more than 0.4 mm spacing. This equates to a void spacing factor of 0.2 mm (the maximum distance from any point in the capillary system to the surface of the nearest air bubble). To achieve this at a 5% air content, most of the air bubbles must be between 0.3 and 0.1 mm diameter. This will give a specific surface to the bubble of between 20 and 40 mm^{-1}. If one air bubble of 10 mm diameter gave 5% air in a concrete mix, it would take 300 000 bubbles of 0.1 mm diameter to give a specific surface of 40 mm^{-1} at the same air content. These air voids are still about a hundred times larger than the capillaries.

When the concrete becomes wet, water will be drawn in by capillary suction but any water that enters an air void will prefer to be sucked back into the capillary system. This means that in most situations the air voids remain dry. The exception is when the water enters the concrete under pressure. If the water in the capillaries starts to freeze, the pressure in the capillary system rises but before it reaches a level which would crack the concrete, unfrozen water is forced into the air voids and the pressure is released. The effectiveness of this system is demonstrated in Figure 4.14 where even the minimum air content of 3.5% is enough to dramatically increase the number of freeze–thaw cycles before serious damage occurs to the concrete. This is compared to a high-strength-water reduced concrete, which shows significant distress after only about 60 freeze–thaw cycles.

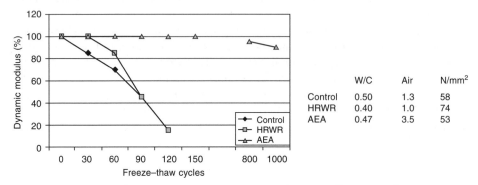

Figure 4.14 ASTM C666 freeze–thaw test on concrete prisms shows the effectiveness of air entrainment.

4.5.3 Air entrainment to reduce bleed

Bleed is usually a function of sand grading and is difficult to overcome. Reducing water content/workability or increasing sand are relatively ineffective and an alternative sand source may not be cost-effective but a small amount of entrained air can make a significant difference.

Bleed can result in sand runs or water lenses under coarse aggregate that show as surface defects. Paste or grout may wash out if there are any holes in shutters. On the surface a weak layer of high w:c paste and latence may form. Bleed is actually a function of sand settlement and may lead to plastic settlement cracking, which forms from the top layer of reinforcement up to the surface and is a serious durability risk.

One to two per cent of additional air can very significantly reduce most bleed problems without having an over-serious effect on strength. Indeed, some plasticizers will put in this level of air as a secondary effect and may be sufficient to deal with the problem. The benefit from the air probably comes from its ability to form air voids that match the gaps in the sand grading, filling the spaces and reducing the opportunity for settlement.

4.5.4 Compaction and cohesion of low-workability mixes

Mixes with very low workability (zero slump), and/or which have a low cement content and are harsh can be very difficult to place and compact. In these mixes, the presence of some entrained air can greatly enhance the compaction under vibration. The air helps to lubricate contact points where the aggregates touch and would otherwise prevent normal flow and compaction. In some cases air may be formed at these points under the action of the vibrator and be lost again when vibration stops. The admixture helps the mix to compact under vibration, closing voids and forming a good surface but the mix stiffens again very quickly when vibration stops, preventing further slumping. This helps to preserve the shape of the unit, especially where a free vertical surface is formed in extruded or instantly demoulded units.

Typical applications include: horizontal slip-forming, extruded slabs, blinding, screeds, semi-dry bricks and blocks.

4.5.5 Bedding mortars and renders

Air is extensively used in bedding mortars and renders as a 'plasticizer' to reduce water content and give a softer feel during trowelling. It also gives some protection against frost action, even at an early age and helps to control strength in richer mixes. The air content is much higher than that used in concrete, being typically in the range 14–20%. Admixtures for this application are covered by EN 934-3.

4.6 Water resisting (waterproofing)

These are defined in EN 934-2 as admixtures that reduce the capillary absorption of water into the hardened concrete. There are, in addition, 'permeability-reducing' admixtures that reduce the passage of water through the concrete under a pressure head. Most products function in one or more of the following ways:

- Reducing the size, number and continuity of the capillary pore structure
- Blocking the capillary pore structure
- Lining the capillaries with a hydrophobic material to prevent water being drawn in by absorption/capillary suction.

All these 'waterproofing' admixtures reduce surface absorption and water permeability of the concrete by acting on the capillary structure of the cement paste. They will not significantly reduce water penetrating through cracks or through poorly compacted concrete which are two of the more common reasons for water leakage through concrete.

Dosage varies widely, depending on the type and the required level of performance. Two per cent is common for the hydrophobic types, but may be 5% or more for the pore blockers. Dosage is often given as a weight or volume per cubic metre rather than on cement weight as their action is largely independent of cement content.

Lowering the free water content of the mix reduces the size and continuity of capillary pores and this can be achieved with water-reducing or high range water-reducing materials. Reducing capillary size alone does not necessarily reduce surface absorption and may even increase it as capillary suction increases with smaller diameter pores. However, the overall pressure permeability of the concrete will be improved due to greater capillary discontinuity.

Pore blocking can be achieved by the addition of very fine unreactive or reactive additions such as silica fume or by the use of insoluble organic polymers such as bitumen introduced as an emulsion. The hydrophobic admixtures are usually derivatives of long-chain fatty acid of which stearate and oleate are most commonly used. These may be supplied in liquid or powder forms. Some admixtures are combinations of two or more of these systems.

4.6.1 Mechanism

Water will be absorbed, even into dense concrete, through the capillary pores. However, if the added mix water can be reduced to a w:c of below about 0.45 (or 160 litres per cubic metre of added water) the capillary system loses much of its continuity and pressure permeability is significantly reduced. The best way of achieving this reduced level of water is to use a water-reducing admixture that can also be used to ensure sufficient workability for full compaction and reduce shrinkage cracking.

The pore-blocking admixtures are based on very fine reactive or unreactive fillers or insoluble polymer emulsions, which have particle sizes of around 0.1 microns and are small enough to get into the capillaries during the early stages of hydration and physically block them.

The hydrophobic admixtures are usually designed to be soluble as an admixture but react with the calcium of the fresh cement to form an insoluble material which adsorbs onto the surfaces of the capillaries. Once the capillary dries out, the hydrophobic layer prevents water re-entering the capillary by suction but resistance is limited and depends upon the head of water involved, the quality of the concrete and the effectiveness of the admixture.

4.6.2 Admixture selection

Hydrophobic 'water-resisting admixtures are generally effective:

- against rain
- against surface water
- against low-pressure heads in structures
- water ingress in tidal and splash zone
- reducing chloride diffusion by preventing build-up of absorbed chloride at the surface.

They are not effective against water with a continuous pressure head.

Pore blockers are best if used in combination with a high range water reducer. They are effective in reducing water penetration under a pressure head of several atmospheres and permeability coefficient can be improved from typically 10^{-10} to better than 10^{-13} m/s.

Most waterproofing admixtures tend to slightly reduce the strength of concrete, either due to some increased air entrainment or to reduced cement hydration. Integral water reduction is often enough to offset this strength loss but some admixtures, especially the liquid hydrophobic types, may need the separate addition of a water reducer.

4.7 Corrosion-inhibiting admixtures

Corrosion-inhibiting admixtures are not currently covered by EN 934 but are effective after the concrete has hardened and give a long-term increase in the passivation state of steel reinforcement and other embedded steel in concrete structures. The three most common generic types of corrosion inhibiting admixture are:

- Calcium nitrite (normally contains a residual amount of calcium nitrate)

- Amino alcohols
- Amino alcohols blended with inorganic inhibitors

The dosage of corrosion inhibitors is usually dependent upon the client's expected serviceable life of the structure and on a range of factors that affect the durability of concrete. These include cement type, water-to-cement ratio, cover concrete to the steel, ambient temperature and the expected level of exposure to chlorides. The typical dosage range for a 30% solution of calcium nitrite is 10–30 litres/m^3 but is more usually used between 10 and 20 litres/m^3. The dosage of amino alcohol-based corrosion inhibitors is usually between 3 and 4 volume per cent by weight of cement. Both types can be used with other admixture types and their use with a high range water-reducing admixture is usually recommended in order to ensure the quality and durability of the base concrete.

4.7.1 Mechanism

The most common cause of reinforcement corrosion is pitting corrosion due to the ingress of chloride ions through the covering concrete and subsequent diffusion down to the embedded steel. When the amount of chloride at the steel surface reaches a critical level, the passivation of the steel breaks down and corrosion starts. Carbonation of the concrete leads to a lowering of the alkalinity around the steel and when the pH drops below a critical level this causes a loss of passivation that results in general reinforcement corrosion.

Corrosion inhibitors increase the 'passivation state' or 'corrosion threshold' of the steel so that more chloride must be present at the steel surface before corrosion can start. Although corrosion inhibitors can raise the corrosion threshold, they are not an alternative to using impermeable, durable concrete and are not cost-effective unless used in a high-quality mix.

The mechanism by which corrosion inhibitors operate is dependent on their chemical nature. Calcium nitrite-based corrosion inhibitors convert the passive layer on the steel surface into a more stable and less reactive state. When the chloride ions reach the layer, no reaction occurs – the steel is in a passive state. It is the anodic corrosion sites on the steel that are protected against the chloride attack and for this reason nitrites are called anodic inhibitors. It is important that sufficient nitrite is present to counter the chloride ions. The dosage of the admixture is therefore based on the predicted level of chloride at the steel over the design life of the structure.

Amino alcohol-based corrosion inhibitors coat the metal surface with a monomolecular layer that keeps the chloride ions away from the embedded steel. They also inhibit the reaction of oxygen and water at the cathodic sites on the steel, which is an essential part of the corrosion process. As a result amino alcohols can be regarded as both anodic and cathodic inhibitors.

4.7.2 Use of corrosion inhibitors

Corrosion inhibitors can significantly reduce maintenance costs of reinforced concrete structures throughout a typical service life of 40–60 years. Structures especially at risk are those exposed to a maritime environment or other situations where chloride penetration of the concrete is likely. Such structures include jetties, wharves, mooring dolphins and sea walls. Highway structures can be affected by the application of de-icing salts during winter months, as can multi-storey car parks where salt-laden water drips off cars and evaporates on the floor slab.

The concrete should be designed with a low diffusion coefficient and sufficient cover to slow the rate of chloride reaching the steel, the corrosion inhibitor can then extend the time to onset of corrosion as indicated by the following example:

Concrete spec.
PC 400 kg/m^3 w:c 0.40
Reinforcement at 40 mm
Chloride diffusion coeff. 1×10^{-12} m^2/s
No corrosion inhibitor
Chloride corrosion threshold 0.4% on cement
Threshold reached after 20 years
Calcium nitrite added at 10 litres/m^3 (30% soln)
Chloride corrosion threshold raised to 0.8% on cement
Threshold reached after 50 years

In this example, the inhibitor has not reduced the amount of chloride that reaches the steel but has increased the amount of chloride that must be present at the steel surface before corrosion is likely to start.

4.8 Shrinkage-reducing admixtures

Shrinkage-reducing admixtures can significantly reduce both the early and long-term drying shrinkage of hardened concrete. This is achieved by treating the 'cause' of drying shrinkage within the capillaries and pores of the cement paste as water is lost. This type of admixture should not be confused with shrinkage-compensating materials which are normally added at above 5% on cement and function by creating an expansive reaction within the cement paste to treat the 'effects' of drying shrinkage.

Shrinkage-reducing admixtures are mainly based on glycol ether derivatives. These organic liquids are totally different from most other admixtures, which are water-based solutions. Shrinkage reducing admixtures are normally 100% active liquids and are water-soluble. They have a characteristic odour and a specific gravity of less than 1.00. The dosage is largely independent of the cement content of the concrete and is typically in the range 5–7 litres/m^3.

4.8.1 Mechanism

When excess water begins to evaporate from the concrete's surface after placing, compacting, finishing and curing, an air/water interface or 'meniscus' is set up within the capillaries of the cement paste. Because water has a very high surface tension, this causes a stress to be exerted on the internal walls of the capillaries where the meniscus has formed. This stress is in the form of an inward-pulling force that tends to close up the capillary. Thus the volume of the capillary is reduced, leading to shrinkage of the cement paste around the aggregates and an overall reduction in volume of the concrete.

The shrinkage-reducing admixtures operate by interfering with the surface chemistry of the air/water interface within the capillary, reducing surface tension effects and consequently reducing the shrinkage as water evaporates from within the concrete. They may also change the microstructure of the hydrated cement in a way that increases the mechanical stability of the capillaries.

4.8.2 Use of shrinkage-reducing admixtures

Shrinkage-reducing admixtures can be used in situations where shrinkage cracking could lead to durability problems or where large numbers of shrinkage joints are undesirable for economic or technical reasons. In floor slabs, the joint spacing can be increased due to the reduced movement of the concrete during drying. The risk of the slab curling at joints and/or edges is also significantly reduced. Where new concrete is used to strengthen or repair existing structures, shrinkage-reducing admixtures can reduce the risk of cracking in what can be a highly restrained environment.

4.9 Anti-washout/underwater admixtures

Underwater concrete anti-washout admixtures are water-soluble organic polymers which increase the cohesion of the concrete in a way that significantly reduces the washing out of the finer particles, i.e. cementitious material and sand from fresh concrete when it is placed under water. They are often used in conjunction with superplasticizers to produce flowing self-levelling concrete to aid placing and compaction under water.

Anti-washout admixtures were developed to provide a higher integrity of concrete placed under water and to reduce the impact that the washed-out material can have on the marine environment. They are suitable for use in deep underwater placement, in inter-tidal zones and in the splash zone and other situations where water movement may result in the cement and other fine material being washed out.

The dosage typically ranges from about 0.3–1.0% depending on manufacturer and on the degree of washout resistance required. It is generally preferable to use the lowest dose that is consistent with the degree of washout resistance needed. The anti-washout admixtures are usually powder-based products, which makes dispensing difficult. Liquid products are available but may be less effective or also have storage/dispensing problems.

4.9.1 Mechanism

The anti-washout admixtures produce an open-branched polymer network in the water, which locks together the mix water, cement and sand, reducing the tendency for dilution with external water during and after placing. The cohesion of the mix is increased, reducing workability and flow. As underwater concrete needs to have high flow and be self-compacting, superplasticizers are needed to recover the lost workability.

4.9.2 Use

Anti-washout admixtures work on the mix water but are more effective if the base mix is cohesive. Cement contents should be at least $400 \, kg/m^3$ but this can include blended cement. The increased cohesion provided by very fine binders such as silica fume have been found particularly effective in some applications. Sand contents should also be high, typically 45% or higher, and need to have a uniform medium or fine grading.

The concrete can be placed by skip, pump or tremie and at higher admixture dosages may tolerate some freefall. However, care is needed not to allow the concrete to fall through a water-filled pump line or tremie pipe, as the turbulent flow produced will cause the mix to segregate.

There is no currently recognized British Standard to measure the performance of anti-washout admixtures. Most comparisons are based on plunge tests where a known mass of concrete is dropped through water a number of times in a wire cage and the degree of washout determined as percentage loss in mass.

4.10 Pumping aids

Modern pumps are able to cope with most concrete mixes but difficulties can be experienced in the following situations:

- Concrete that is produced with sand that is poorly graded or coarse aggregate that is elongated or flaky. This can lead to blockages or segregation within the pipeline and can usually be helped by a low-level air entrainment.
- Pumping for long distances or to high levels can lead to very high pump pressures and the need for improved lubrication within the lines. Pumping aids can also provide better workability retention for long pumping distances and can be particularly effective if there is likely to be delays or breaks in the pumping when segregation or stiffening in the line could prevent a restart.
- Lightweight aggregates are often porous, with water absorption much higher than normal aggregate. Pumping these concretes presents problems, particularly if the aggregates are not pre-soaked and fully saturated. The pump pressure drives water into the aggregate causing the mix to dry out and block the lines. Specifically designed pumping aids are available to assist in the pumping of lightweight aggregate concretes by reducing the amount of water which becomes absorbed and allowing drier aggregate to be used.

The principal chemicals used are long-chain polymers of various types or surfactants similar to those used for air entrainment. The surfactants are usually blended with water-reducing admixtures, which allow slightly increased workability without loss of properties. The long-chain polymer gives cohesion to the mix and prevents bleeding and water loss. They help to maintain a layer of cement paste round the coarse aggregate particles and to produce a slippery layer at the concrete/pipe interface. Surfactants act by increasing paste volume and reducing aggregate interlock.

4.11 Sprayed concrete admixtures

Sprayed concrete is pumped to the point of application and then pneumatically projected into place at high velocity. The applications are frequently vertical or overhead and this requires rapid stiffening if slumping or loss by concrete detaching from the substrate under its own weight is to be avoided. In tunnelling applications, sprayed concrete is often used to provide early structural support and this requires early strength development as well as very rapid stiffening.

Admixtures can be used in the fresh concrete to give stability and hydration control prior to spraying. Then by addition of an accelerating admixture at the spray nozzle, the rheology and setting of the concrete are controlled to ensure a satisfactory build-up on the substrate with a minimum of unbonded material causing rebound.

There are two spraying processes:

- The dry process where the mix water and an accelerator are added to a dry mortar mix at the spray nozzle.
- The wet process where the mortar or concrete is pre-mixed with a stabilizer/retarder prior to pumping to the nozzle where a liquid accelerator is added.

The wet process has become the method of choice in recent times as it minimizes dust emissions and gives more controlled and consistent concrete.

Accelerators are of two types:

- Alkaline, which give the quickest set, often less than 1 minute, and very early strength development but presents a safety hazard to the applicators and tend to give low later age strengths.
- Alkali-free, which can take tens of minutes to set but is safer for the applicator and has little effect on later age strength.

Sprayed concrete admixtures are often affected by the cement chemistry and their use requires technical and practical expertise. For these reasons, their use should be left to specialist applicators.

4.12 Foamed concrete and CLSM

Low-strength, low-density concrete has a number of applications including backfilling of utility trenches, filling old tunnels, sewers, tanks and other voids in or beneath structures. CLSM, controlled low-strength material is a sand-rich concrete with a high dose of a powerful air-entraining admixture. This gives air contents up to about 25% with densities down to about 1800 kg/m^3 and strengths of about 4 N/mm^2. To get lower densities, the air must be added in the form of a foam, which is produced from a surfactant admixture in a foam generator. This is then blended with a sand cement mortar or with a cement grout to give densities down to 600 kg/m^3 or lower. Strength depends on the cement content and density but is typically in the range 0.5–10 N/mm^2.

The foamed concrete/CLSM is very fluid and essentially self-levelling. It can flow long distances and completely fill difficult shaped voids. It is free of bleed or settlement and as it sets, the heat of hydration expands the air, tightening the mortar into the void so that no gaps remain. Its low strength and high level of air entrainment make it easy to remove at some future time. It has many other advantages related to specific applications.

4.13 Other concrete admixtures

4.13.1 Polymer dispersions

Polymer dispersion admixtures are aqueous dispersions of elastomeric/thermoplastic synthetic polymers that will form a continuous film under the conditions of use when sufficient water is lost from the system.

Typical dose rate is 10–30 litres of a 50% solids dispersion per 100 kg of cement. They:

- Reduce the permeability of concrete, improving durability.
- Give a more reliable bond to base concrete and to reinforcing steel.
- Increase flexibility and toughness and resistance to abrasion/dusting.

For economic reasons applications are usually restricted to thin section concrete, e.g. concrete up to about 75 mm in thickness. Polymer dispersions are also widely used in screeds, mortars, renders, and grouts.

4.13.2 Pre-cast, semi-dry admixtures

Admixtures are used in the manufacture of 'dry' or 'semi-dry' vibrated and pressed concrete products such as paving and masonry blocks, bricks, flags and architectural masonry to assist compaction, disperse cement, colour and fine aggregate particles, and help to control primary and secondary lime staining (efflorescence). Admixtures can be specific to only one of these functions or can be multi-purpose.

Semi-dry concrete mixes with very low water contents (typically 6–9%) are made easier to compact by admixtures that reduce particle attraction and surface tension, making the mix more susceptible to vibration and pressure, thereby increasing compacted density. The reduced particle attraction prevents agglomeration and disperses cement and colour particles.

The plasticizing action of these admixtures is only effective under vibration and must not lead to an increase in workability as this could cause the mix to deform before setting.

4.13.3 Truck washwater admixtures

Washwater admixtures are designed to overcome the problem of disposal of washwater following the cleaning of the inside of truck mixer drums at the end of the day. The solution to this is to recycle the washwater into the following day's concrete by the use of a truck washwater admixture.

The washwater admixture stabilizes the concrete washwater in the drum of a readymixed truck on an overnight or over-weekend basis, preventing the cement from hydrating and solid material from forming a layer of hard settlement at the bottom of the drum or on the blades. The washings are then incorporated as water and filler in the production of the next day's initial load.

Dosage varies between manufacturers depending on the materials used and the period before fresh concrete is to be mixed but is typically between 1 and 3 litres for a 6 m^3 truck. The driver operates a lance that discharges the admixture and water at high pressure to wash the residue to the bottom of the drum. The dispenser is programmed to deliver exactly 200 litres of water into the drum. The drum is rotated to finally wash off the residues and is then left static until the next day (or over a weekend if appropriately dosed). The only alteration to the next mix is to reduce the mix water by 200 litres.

4.14 Mortar admixtures

Although any combination of sand and cement can be thought of as a mortar, mortar admixtures are usually intended for use in mixes of 3:1 or leaner for use in bedding of masonry or as renders. At this sand content, the plasticizing admixtures used for concrete are relatively ineffective but air entrainment lubricates the interlock between sand particles giving large reductions in water content. For this reason, air entrainers are sold as and usually called mortar plasticizers.

Almost any surfactant, including washing-up liquid, will have this effect but controlling the air content for different types of mix, mixer, mixing times, and then providing stability against air loss up to the time of hardening is not so simple. The admixtures supplied are often multi-component blends of surfactants and stabilizers designed to optimize these requirements.

A good plasticizer provides additional benefits by reducing density, improving the trowelling characteristics and providing freeze–thaw resistance from soon after the mix has hardened.

Masonry mortar is often provided by readymix in the form of a 'Ready-to-Use Mortar', which has been retarded to allow use for typically up to 36 hours after delivery. When placed between absorbent masonry units, some water and retarder are sucked out and this initiates the hardening process of the mortar between the units even though the unused part of the mortar delivery is still plastic and usable. If the masonry is not absorbent, 'Ready-to-Use Mortar' should be used be used with care and retardation restricted to 8 hours' working life. In the case of renders, working life should always be restricted to 8 hours' working life or a soft, dusting surface is likely to be obtained. Air entrainers for 'Ready-to-Use Mortar' are particularly critical, as any air loss over the 36-hour working life will result in a serious loss of consistency.

Other admixtures can be used with mortars, the most common being:

* Waterproofing admixtures to reduce moisture movement through renders.
* Polymer dispersions, to aid bonding, increase flexibility and give some water repellency.
* Water-retaining admixtures to reduce the suction of water from the mortar into the masonry unit and enhance cohesion/reduce bleed.

Only mortar plasticizers and retarders are currently covered by standards BS 4887 and EN 934-3.

4.15 Grout admixtures

Grout normally refers to a fluid mix of cement and water but may also include sand up to about 2:1. Most conventional concrete admixtures are effective in these mixes, especially superplasticizers, which can give large water reductions while retaining a very fluid mix. Air entrainers are not very efficient in cement-rich mixes.

EN 934-4 refers to grout admixtures mainly with respect to grouts for filling the ducts in post-tensioned structures. The standard requires high fluidity at a w:c below 0.42, fluidity retention reduced bleed and shrinkage while the grout is still fluid and also covers the use of expanding admixtures to overcome early age volume loss. These admixtures are normally gas-expanding systems that release hydrogen or nitrogen over a period while the grout is fluid. This bulks out the grout, helping it to expand and fill any voids caused by settlement. Bleed water can be forced out of a venting valve at the high point of the duct. The stress induced by these admixtures is limited because of the compressibility of the gas. Expansive cement should not be used for this application as the volume expansion occurs after the grout has hardened, which is too late to be effective and creates high stresses that can crack the concrete element.

Anti-bleed admixtures are also available. These are mainly used in tall vertical ducts where water separation can result in the formation of large water lenses. The problem is particularly bad when 7-wire cables are used as water can wick up the inside of the strands. The anti-bleed admixtures not only prevent cement settlement but also seal the gaps between wires in the strands, reducing the wick effect.

4.16 Admixture supply

4.16.1 Suppliers

Admixtures should only be accepted from a reputable company able to demonstrate that they operate a third-party certified quality management system complying with ISO 9001 or 9002. The company must be able to show that their products have been properly tested and comply with the appropriate standard where this is applicable.

Suppliers should be able to provide a data sheet outlining the properties and performance of the product. It should give dosage information including the effects of overdosing, any secondary effects such as retardation or air entrainment, instructions for storage, handling and use, compatibility with other admixtures and cement types, and any other information on situations where there may be restrictions or special requirements on use. The appearance, specific gravity, solids, chloride and alkali contents of the admixture should be clearly stated. A sheet covering health and safety data (MSDS) must be provided for the purchaser and the user at the batching plant. The supplier should also provide a reasonable level of local technical backup for the products.

In the United Kingdom, the CAA (Cement Admixtures Association) requires that all the provisions detailed above are met by its members and in Europe, national associations who are members of EFCA (European Federation of Concrete Admixture Associations) also meet these requirements.

4.16.2 Storage

Admixtures are usually stored in bulk, in plastic or metal tanks or in barrels. Modern bulk plastic storage tanks are double-skinned in order to minimize the risk of leakage due to a skin splitting. If double-skinned tanks are not used, the admixture storage area should be bunded in order to prevent any spillage entering the water system.

Admixtures should be protected from freezing, all tanks and pipes should be clearly marked with their contents and the shelf life of the product should be indicated.

Admixture containers should never be left with open tops due to the risk of contamination. Rain can also enter, diluting the product and reducing its effectivness.

If an admixture exceeds its storage life or is no longer required, it must be disposed of through an approved contractor.

4.16.3 Dispensers

The dosage of most admixtures is only in the range 0.2–2.0% volume for weight on cement. This means that relatively small changes in actual volume of admixture dispensed can have a large effect on dosage as a percentage of cement and thus significantly affect the quality and properties of the concrete. For this reason, it is important that admixtures are added to the concrete through an accurate, frequently calibrated dispenser.

The dispenser should include an automatic water-flushing system to clean it and ensure that all admixture is fully washed into the mixer. This is essential where several admixtures are dispensed through the same dispenser or cross-contamination can occur, potentially leading to unwanted effects.

There are three main types of admixture dispenser used in the UK:

- Volumetric measurement
- Weighing
- Volumetric metering

All three types of dispenser have a successful history of use but each has its own strengths and weaknesses. The admixture supplier should be able to give advice on the most appropriate type.

4.16.4 Time of admixture addition

For optimum performance most admixtures should be added at the end of the batching process, with some of the mixing water. Liquid admixtures should never be added directly onto dry cement. The timing of the addition affects the performance of the admixture, therefore it is important that the admixture is always added at the same point in the batching process to ensure consistent performance. Particular attention to the mixing is needed to ensure that all the admixture is uniformly distributed through the entire mix.

Certain admixtures can be added on-site, e.g. some superplasticizers and foaming admixtures, and it is important that in this case accurate and safe dispensing methods are available. All admixture additions should be through a calibrated, metered pump and be carried out under supervision and to a prepared plan agreed by all parties.

4.17 Health and safety

The major admixture companies all supply safety data sheets for their products and these should be carefully checked and any advice on handling passed to those working with the products. Most admixtures are water-based and are non-hazardous or, at worst, irritants. A small group have a high pH and are therefore harmful/corrosive: this particularly applies to some air-entrainers, waterproofers and shotcrete accelerators. A very small number; mainly from the accelerators and corrosion inhibitors, may be toxic if ingested.

Some organic admixtures have a high oxygen demand under biological breakdown that, during disposal or in the event of a major spillage, may cause problems for fish in rivers and in processing of effluent in water-treatment plants. Any such problem should be notified to the appropriate authorities immediately it occurs.

Further reading

A Guide to the Selection of Admixtures for Concrete	Concrete Society TR 18
Cement Admixtures: Use and Application	CAA
Chemical Admixtures for Concrete	Rixom
Applications of Admixtures in Concrete	Paillere
Concrete Admixtures Handbook	Ramachandran
Superplasticisers & Other Chemical Admixtures	CANMET/ACI Rome 97 ACI SP-173
Lea's Chemistry of Cement & Concrete	Hewlett
Water-reducing admixtures in concrete	BRE Information Paper IP 15/00

The UK Cement Admixtures Association also produces a number of guidance papers (further details on their website, admixtures.org.uk.

PART 4

Aggregates

5

Geology, aggregates and classification

Alan Poole and Ian Sims

5.1 Introduction

Natural rock in the form of aggregate particles typically makes up between 70 per cent and 80 per cent of the volume of a normal concrete. The word 'aggregate' is a familiar one, perhaps because rock fragments or gravels are one of the most important common bulk materials in general use, particularly in the building industry. Natural sand, gravel and crushed rock undoubtedly form a major and fundamental part of concrete and mortars, but they are also very important constituents of road-making asphalts and macadams, and are used extensively as filters and drainage layers, and as railway ballast.

Particles of natural rock are by far the commonest form of aggregate, but recycled crushed concrete and manufactured materials such as furnace slags and expanded clay, shale or slate pellets are also used to a more limited extent. The aggregate as a material must be strong, durable and inert to give satisfactory performance, and the sizes of the constituent particles must be appropriate for the intended application. They are described as coarse aggregate, in the UK if particles are retained on a screen with 5 mm apertures, or 4 mm apertures, if recent European standard specifications (CEN TC154SC2, EN 12620, 2002) are followed. They are described as fine aggregate, or sand if they pass through them. Aggregates are normally separated into size fractions by the use of a series of different sized screens, but within any fraction there will be a grading of sizes from those particles that can just pass the larger screen to the ones that are just fractionally too large to pass the smaller screen. Where a wider range of sizes is necessary for, say, a concrete, several size fractions may be selected and mixed together to give the appropriate overall grading.

5.2 Fundamentals

The rocks of the Earth's crust such as those we see on the surface today were formed at various times in the geological past and in a variety of ways. The geological time scale has been developed as the result of many research studies all starting from the pioneering work of William Smith in 1815 (Winchester, 2001). A modern consensus of the geological time scale is shown in Table 5.1.

Table 5.1 The Geological Column, based on information from Eagar and Nudds (2001) and other sources

Age (Ma BP)*	Epoch	Period	Era	Eon
	Holocene (recent)	Quaternary		
0.01				
	Pleistocene			
2.5				
	Pliocene	Neogene	Cenozoic	
5.2				
	Miocene			
23.5		Tertiary		
	Oligocene			
35.5		Paleogene		
	Eocene			
56.5				
	Palaeocene			
65				
		Cretaceous		
146				
		Jurassic	Mesozoic	
205				
		Triassic		
251				
		Permian		
290				
		Carboniferous		
353				
		Devonian	Palaeozoic	
409				
		Silurian		
439				
		Ordovician		
510				
		Cambrian		
540				
		Precambrian		Proterozoic
2500				
		Precambrian		Archaean
4500			Formation of the crust	
4600			Origin of the Earth	

(Numerous epochs worldwide)

Phanerozoic ('evident' life, i.e. many fossils in sedimentary rocks)

*MaBP: Millions years before the present.

The geological age of a source of rock for aggregate is usually given along with its name. This allows the rock to be placed in its correct geological setting and history. Also, with many rocks, particularly sediments, the age may provide a clue to its properties. Although there are many exceptions, geologically older rocks are likely to be better lithified, that is, solidified, and consequently are likely to be harder and more durable than younger ones of the same type. As an example, the differences in the physical properties of Cretaceous Chalk and Carboniferous Limestone are obvious, though both are limestones. A major factor accounting for the difference is that the chalk is about 300 million years younger than the Carboniferous Limestone.

All natural rock materials are composed of aggregations of one or more mineral constituents. A mineral is defined as a natural inorganic substance with a definite chemical composition, atomic structure and physical properties (Kirkaldy, 1954). Often the individual mineral crystals or grains are very small, with some rocks the minerals are so small that they cannot be seen without the aid of a magnifying glass, but in others, notably certain granites, the individual mineral crystals can be several centimetres long.

The chemical elements from which these minerals are formed are of course those present in the Earth's crust. Perhaps surprisingly, eight chemical elements account for more than 99 per cent by mass of all crustal rocks as is shown in Table 5.2. These eight elements combine together in various ways to form the common rock minerals and may be summarized in tabular form as in Table 5.3.

Table 5.2 The abundance of chemical elements in the Earth's crustal rocks by mass. (After Carmichael, 1982)

Element	% mass
Oxygen	46.4
Silicon	28.2
Aluminium	8.2
Iron	5.6
Calcium	4.2
Magnesium	2.3
Sodium	2.4
Potassium	2.1
Others	0.6
Total	100

Table 5.3 The common rock-forming minerals

Minerals			RD*	H**
Silica	Quartz	SiO_2	2.65	7
Silicates				
Primary	Felsic (light)	Al, Ca, Na, K, Si	2.57–2.74	6.65
	Mafic (dark)	Mg, Fe, Si	2.8–4.0	5–5.7
Secondary	Clay minerals	Al, Si, K	2.2–2.4	2–3
Carbonates	Calcite	$CaCO_3$	2.71	3

* RD The relative density range of the minerals
** H This is the typical hardness of the mineral(s) based on the Mohs' Scale of mineral scratch hardness where talc is given a hardness of 1, quartz 7 and diamond 10. (The fingernail will scratch up to $2^1/_2$ and a penknife blade $6^1/_2$)

Rocks may be divided into three broad groups. The first, **igneous** rocks, are formed when molten rock material (called magma) is generated below or within the Earth's crust and crystallizes as solid rock as it cools down, either on the surface as a lava or within the Earth's crust as an intrusion. Since igneous rocks are intruded into pre-existing rocks in various ways and are now seen at the Earth's surface as a result of erosion, the series of descriptive terms used to describe their occurrences are best illustrated pictorially as in Figure 5.1.

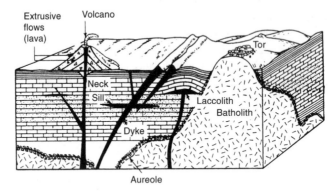

Figure 5.1 An idealized block diagram showing common forms and associations of intrusive and extrusive igneous rocks (after Fookes, 1989).

The second group are **sedimentary** rocks which are formed by the accumulation of fragments of pre-existing rocks resulting from processes of erosion, organic debris such as shell fragments or plant material. These are the detrital sedimentary rocks, or alternatively, they may be formed as a chemical precipitate from oversaturated sea, or ground waters, the chemical and biochemical sedimentary rocks.

The third group, **metamorphic** rocks, are formed from pre-existing rocks of any type, sedimentary or igneous, which have then been subjected to long periods of increased temperature and/or pressure within the crust. Depending on the severity and the time rocks are subjected to these high temperatures and pressures they undergo progressive changes ('metamorphism') resulting in new minerals being formed and modifications to their original appearance (McLean and Gribble, 1990).

5.3 Geological classification of rocks

Rocks are usually identified by their rock name and sometimes also the geological period in which they were formed. Geologists have devised numerous rock name classifications usually based on the sizes and types of mineral particles or crystals that they contain. Some are very elaborate and detailed, but for general purposes and as a first step in aggregate identification, simple systems for igneous, sedimentary and metamorphic rocks such as are illustrated in Tables 5.4–5.6 should be adequate.

It was noted above that metamorphic rocks are formed from pre-existing rocks by the action of heat and/or pressure extending over a variable period of geological time. These are consequently difficult to classify in detail because of their variety and the range of

Table 5.4 Simplified igneous rock classification

		Composition			Occurrence
		Acid	Intermediate	Basic	Ultrabasic	
	Silica%	>65	55–65	45–55	<45	
	Quartz content	Much quartz	Little quartz	Almost no quartz	No quartz	
	Colour	Pale		Dark	Very dark	
	Density	2.4–2.7	2.7–2.8	2.9–3.0	3.0	
Grain size	Coarse-grained (>2 mm)	Granite	Diorite	Gabbro	Peridotite	Large intrusions
	Medium-grained (2–0.5 mm)	Micro-granite	Micro-diorite	Dolerite		Small intrusions Sills Dykes
	Fine-grained (<0.5 mm)	Rhyolite	Andesite	Basalt		Lava flows and small dykes

Table 5.5 Simplified sedimentary rock classification

Main mode of origin	Dominant chemical constituents	Rock name group	Examples
Detrital	Silica and silicates	Terrigenous	Conglomerate Sandstone Mudstone Shale
Chemical and biochemical	Carbonates	Carbonate rocks	Limestone Chalk
	Chlorides and sulfates	Evaporites	Rock salt, Gypsum
	Various forms of silica	Siliceous rocks	Chert, Flint
	Ferruginous minerals	Ironstones and iron-rich rocks	Sedimentary ironstones
	Phosphates	Phosphorites	Phosphorites
Biological	Carbon and organic residues	Carbonaceous rocks	Coal Sapropelites
Deposits from volcanic sources	Silicates	Volcaniclastic rocks	Agglomerates Tuffites Ignimbrites

Table 5.6 Simplified metamorphic rock classification

	Regional metamorphic rock groups		
	Low grade	Moderate grade	High grade
Grain size	<<1 mm	1–3 mm	>3 mm
Temperature	<300°C	300–550°C	>550°C
Pressure	<3 kb	3–5 kb	>5 kb
Rock fabric	Slaty cleavage	Foliation	Discontinuous banding
Rock types	Slates Phyllites	Schists of all types	Gneisses Migmatites Granulites

possible metamorphic conditions. Table 5.6 divides the general classification into three, but if the effects of the high metamorphic temperatures and pressures are simplified to two, namely, **high grade** and **low grade** (effects often associated with how deeply the rocks are buried in the Earth's crust, since typically both temperature and pressure rise with depth) and related to the original rock types, then a much simplified classification appropriate to possible aggregate materials can be devised as indicated by Table 5.7.

Table 5.7 A generalized classification of metamorphic rocks with special reference to rocks potentially suitable for aggregate

Original rock	Metamorphic environment – temperature and pressure conditions	
	Low grade – shallow burial	Medium and high grade – deep burial
Clay, shale and volcanic tuff clayey sandstone	Foliated platy rocks (effects of pressure paramount) slate greywacke-sandstone*†	phyllite, schist gneiss* quartz-mica schist, fine-grained gneiss calcerous schist
	slaty marble*†	
Clayey limestone granite,* quartzitic tuff	shered granite, slate	granite-gneiss*, quartz-mica schist
Basalt,* basic volcanic tuff	green (chlorite) schist	amphibolite (amphibole schist) hornblende gneiss*
	Non-foliated, massive rocks (effects of increased temperature paramount)	
Any parent rock		hornfels* (some recrystallization but often original features still present)
Quartzose sandstone*	quartzitic sandstone*	quartzite (includes mica if impure) psammite*, granulite* (essentially quartz, feldspar + some mica)
Limestone* and dolomite*	marble* (if impure, contains a wide range of calcium and magnesium silicates with increasing metamorphic grade)	

*Rocks most likely to be useful for aggregate
†Some rocks containing clay minerals may not perform satisfactorily as aggregate in some applications

The classification of the most common sedimentary rocks, that is, the **clastic** (or detrital) rocks and the **limestones**, also require further elaboration. Clastic sediments are named on the basis of particle grain size and whether they are unconsolidated (sands and gravels) or have been consolidated by lithification (hardened, or indurated). This classification is given in Table 5.8.

Limestone, although essentially simple, in that all limestones are composed mainly of calcium carbonate ($CaCO_3$) prove difficult to classify in detail, because when impure they can fall into a range of sedimentary compositions, with a claystone or sandstone at one extreme of the classification, and pure limestone at the other, as is illustrated in Table 5.9.

A limestone can also contain a proportion of dolomite, which is a calcium–magnesium carbonate mineral ($CaMg(CO_3)_2$). If the rock contains over 90 per cent of mineral dolomite it is called a **dolomite** (named from the locality where it is found in the Tyrolean Alps), if there is more than 50 per cent dolomite it is calcitic dolomite, while less than 50 per cent, it becomes dolomitic limestone.

Table 5.8 The classification of detrital (clastic) sedimentary rocks based on grain size

Grain diameter, mm	Category	Term	Indurated equivalent	Ajectival terms
>2.0	–	Gravel	Conglomerate	Rudaceous
0.60–2.00	Coarse	Sand	Sandstone	
0.20–0.60	Medium			
0.06–0.20	Fine			
				Arenaceous
.020–.060	Coarse	Silt	Siltstone	
.006–.020	Medium			
.002–.006	Fine			
<0.002	–	Clay	Mud-rock	Argillaceous

Table 5.9 A simple classification of carbonate rocks. (After Fookes and Higginbottom, 1975)

	Percentage carbonate					Percentage carbonate
	0					0
Claystone		Siltstone	Sandstone	Conglomerate		
	5					10
Marly claystone						
	20					
Limey marlstone		Calcerous siltstone	Calcerous sandstone	Calcerous conglomerate		
	35					50
Marlstone						
	65					
Clayey marlstone		Silty limestone	Sandy limestone	Conglomeratic limestone		
	80					90
Marly limestone						
	95					
Limestone		Limestone				100
	100%					

Yet another complication arises in that many limestones are composed in part from shell fragments and other clastic particles of calcium carbonate. These are called allochemical constituents of a limestone and are listed in Table 5.10. Limestones may contain one or more of these constituents set in a lithified lime mud (**micrite**), or a more coarse recrystallized calcite (**sparite**).

Table 5.10 The allochemical constituent particles commonly found in limestones. They can be used as prefixes to refine the Folk (1959) classification given in Table 5.11

Component	Prefix for use in Table 5.11	Mode of origin
Fossils	bio-	Skeletal parts of carbonate-secreting organisms
Peloids	pel-	Spherical or angular grains without internal structure, mostly of faecal origin
Ooids	oo-	Spherical, or sub-spherical grains with a concentric internal structure, probably of biochemical origin
Aggregates	clastic-	Aggregates of several carbonate and other particles cemented together
Intraclasts	intra-	Fragment of previously lithified, or partly lithified, sediment incorporated into the new sediment

All surface rocks degrade with time under the effects of the weather, eventually changing to a soil, although some are much more durable than others. Research into the weathering of rock has resulted in a generally accepted sixfold scale of rock weathering, with fresh rock as 1 and soil at 6. This scale is shown in Figure 5.2.

Table 5.11 A classification for limestones after Folk (1959) appropriate for shelly limestones. The prefixes from Table 5.10 are appropriate for other allochems

Low water turbulence ——————————→ High water turbulence

	Over $\frac{2}{3}$ lime mud matrix (Micrite)			Subequal spar & lime mud	Over $\frac{2}{3}$ spar cement (Sparry calcite)		
Representative rock terms	Micrite and dismicrite	Fossiliferous micrite	Sparse biomicrite	Packed biomicrite	Poorly washed biosparite	Unsorted biosparite	Sorted biosparite	Rounded biosparite
Percent Allochems 'grains'	0–1%	1–10%	10–50%	Over 50%		Sorting poor	Sorting good	Round and abraded
Terrigenous analogues	Claystone		Sandy claystone	Clayey or immature sandstone		Submature sandstone	Mature sandstone	Supermature sandstone

■ Lime mud matrix

▨ Sparry calcite cement

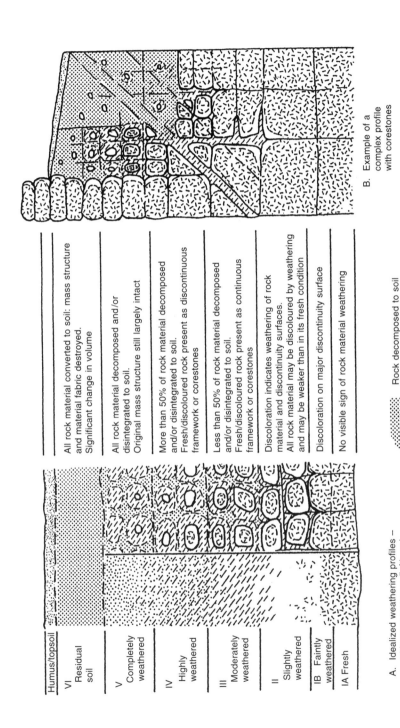

Humus/topsoil

VI Residual soil — All rock material converted to soil: mass structure and material fabric destroyed. Significant change in volume

V Completely weathered — All rock material decomposed and/or disintegrated to soil. Original mass structure still largely intact

IV Highly weathered — More than 50% of rock material decomposed and/or disintegrated to soil. Fresh/discoloured rock present as discontinuous framework or corestones

III Moderately weathered — Less than 50% of rock material decomposed and/or disintegrated to soil. Fresh/discoloured rock present as continuous framework or corestones

II Slightly weathered — Discoloration indicates weathering of rock material and discontinuity surfaces. All rock material may be discoloured by weathering and may be weaker than in its fresh condition

IB Faintly weathered — Discoloration on major discontinuity surface

IA Fresh — No visible sign of rock material weathering

A. Idealized weathering profiles – without corestones (left) and with corestones (right)

B. Example of a complex profile with corestones

Rock decomposed to soil

Weathered/disintegrated rock

Rock discoloured by weathering

Figure 5.2 A pictorial representation of weathering grades. (After Geological Society Engineering Group Working Party Report, 1995.)

As an example, the changes in the physical properties that occur as a granite is increasingly weathered are shown in Table 5.12. This table gives the physical properties in terms of a series of test values appropriate to each of the weathering grades. A similar sequence of change occurs with other rock types as they become weathered. Clearly rock aggregates for concrete need to be both strong and durable. This implies that a rock suitable for use as a concrete aggregate will normally need to be in weathering grades one or two.

5.4 Sources and types of aggregates

Natural rock, sands and gravels are by far the commonest source of aggregate worldwide. Artificial and recycled materials account for only a tiny fraction of the total aggregate produced. In the UK alone, the total aggregate requirement is in excess of 200 million tonnes annually, while in Japan the figure is 420 million tonnes (Fox, 2002).

The production of aggregates from all sources in recent years for the UK is illustrated in graphical form in Figure 5.3 and shows that land-based extraction of sand and gravel aggregate together with crushed rock are the predominant sources, sea-dredged materials account for about 10 per cent of the total while the contribution of manufactured and recycled materials is currently very small. In recent years the demand for aggregate in the UK has not increased, but generally demand closely follows national economic trends.

In the UK, as elsewhere in the world, the highest usage of aggregates are in areas of high population density and where major constructional development is taking place. Hence, in Britain demand for aggregate typically tends to be heaviest in the south-east of England and in areas of the Midlands.

Sources of aggregates fall into the broad categories of **sands and gravels** and **crushed rock**. Sands and gravels are the products of erosion of pre-existing rocks and are usually transported by water, or by ice in formerly glaciated regions. They are typically deposited as relatively thin layers at the foot of mountains, in river valleys or along shorelines. Crushed rock, by contrast, is obtained from rock quarries which imply that appropriate rock must occur at the Earth's surface where a quarry can be developed.

The geology of the British Isles is such that if a line is drawn southwards from the Humber down to Portland Bill, then as a generalization no strong rocks suitable for quarrying for aggregate occur to the east of the line, while to the north and west of it there are many areas where rock quarries could be sited, or have already been developed (the Stone Line, Figure 5.4). As a consequence the aggregate requirements for the densely populated south-east must be met by local sand and gravel deposits, by imported materials or by transporting aggregates long distances from aggregate quarries in the west. Long-haul distances for aggregate are generally undesirable for both environmental and economic reasons (the cost of aggregate can be doubled for every 10–15 km of haul distance).

A more detailed examination of a geological map of the British Isles allows generalizations to be drawn concerning the location and types of hard rock suitable for aggregate. Similarly, consideration of the present and past river systems, the extent of past glaciations across Britain and the extent of the shallow continental shelf, again allows inferences to be drawn concerning sand and gravel resources available within the British Isles. These generalizations are illustrated on the maps in Figure 5.4.

Aggregate is produced from quarries by drilling and blasting the quarry face with explosives to maximize the fragmentation, while minimizing the production of 'fly rock'

Table 5.12 The typical physical properties of granite in the different weathering grades. (After Fookes et al., 1971.)

Mass weathering grade		Uniaxial compressive strength saturated (MN/m³)	Bulk density saturated (g/cm³)	Saturation moisture content (%)	Aggregate impact value (%)	Aggregate impact value modified (%)	Aggregate abrasion value (%)	MgSO₄ soundness value (%) retained	Secondary minerals (%)
I	Fresh	262	2.61	0.11	6	7	3.5	100	6
	Stained rim of block	232	2.62	0.15					
	Whole sample II 90% stained	136	2.58	1.09	8	10	4.7	99.9	10
II	Completely stained II block	105	2.56	1.52	14	16	8.0	99.8	12
	Rock core of III block	46	2.55	1.97					
III–IV	Rock core of IV block	26	2.44	4.13	24	49	17.1	66.6	17
V	Weakly cemented soil (coreable)	5	2.24	10.00	nd	nd	nd	nd	

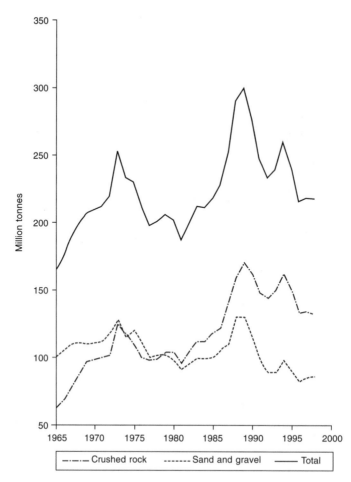

Figure 5.3 The production of aggregate from various sources from the UK over the period 1965–1998. (*Sources:* Quarry Products Association and BGS; Smith and Collis, 2001.)

and vibration. The blasted fragments are then crushed and passed through a series of vibrating screens to produce the particle size fractions required. These sizes depend on local requirements, but a typical example using the European aggregate standard specifications (CEN TC154SC2, EN 12620, 2002) might be 20–14 mm, 14–12.5 mm, 12.5–6.3 mm for coarse aggregate, and 4–2 mm and 2–1 mm for the fine. Usually the rock fragments are reduced to required sizes in a series of steps with reduction ratios of between 10:1 and 5:1. Usually secondary and even tertiary crushing and screening sequences are required with oversize material being recycled through the system and the unusable waste fine dusts removed. Many rock types are suitable for the production of a durable aggregate but the shape of the particles is also an important factor. They ideally should be 'equant', that is, equidimensional in shape. The natural joints in the rock, together with their physical properties, may tend to produce elongate or flaky particles, so that careful choices of crushing plant design must be made in order to counter this tendency.

The aggregates won from natural sand and gravel deposits have several advantages over crushed rock aggregate. Nature, through the processes of erosion and deposition,

Figure 5.4 Generalized maps of the British Isles showing the areas of sands and gravels, and of rocks which may be suitable as aggregates.

tends to concentrate hard durable particles, while at the same time the weaker non-durable material tends to be broken down further to silt and clay which is transported more readily by flowing water than the heavier coarser particles. The processes of deposition are such that the material tends to be at least partially sorted by size and relative density, while abrasion during transportation tends to round the particles to shapes which tend towards spherical. Rounded particles make for a more workable concrete than very angular crushed rock fragments which may be described as giving a 'harsh' mix for concrete and may make its compaction difficult. However, because sand and gravel deposits are usually built up over a period of geological time both the sources from which the materials are derived and the energy of the erosive processes may change on a seasonal, or on a longer-term basis as erosion, or tectonic uplift alters the source area. Examples might be modification to the rainfall pattern, or changes to the river catchment area over time. Such changes will lead to changes in the quantities, sizes and distribution of the sand and gravel particles deposited. Consequently most gravel resources are characterized by rapid and local variations in size distributions and degree of sorting with layers and lenses of coarser and finer materials, which may be interleaved with each other on a scale of a few metres such as is shown in Figure 5.5. The proportion of silt and clay in a deposit which is currently considered waste and must be discarded, together with the proportion of over-size material which would require crushing to reduce its size before it can be used as aggregate, are important economic factors to be considered in the development of a sand and gravel resource.

Another advantage of sand and gravel reserves is that they are usually loose unconsolidated materials which can be dug directly with a face shovel or a similar mechanical excavator. The raw material then only has to be washed to remove clay and silt and screened to the appropriate aggregate size fractions. If oversize particles are present and are crushed down to aggregate sizes the crushed particles are usually blended in with the natural more rounded gravel to avoid the aggregate being too 'harsh'.

Sources of sand and gravel aggregate may be grouped under four headings as indicated in Table 5.13.

It should of course be recognized that although many sand and gravel deposits are of relatively recent geological origin, ancient river systems, glacial deposits and near-shore deposits of earlier geological periods are sometimes worked as sources of sand and gravel.

5.4.1 Temperate fluvial environments

River terrace deposits are perhaps the commonest of the alluvial deposits. These materials are eroded from the upper reaches of rivers, carried downstream by the river, particularly at times of flood when it is most energetic, and deposited in the lower reaches typically on a floodplain where the rate of flow is checked and the river can no longer transport its burden of sediment. This sedimentary material builds up in thickness over time. Most rivers have a history that will extend back through the Quaternary Period with its reduced sea levels during the glacial periods, and often well into the Tertiary Period. Such rivers will have had periods of rapid cutting down through the earlier alluvial deposits to adjust to reduced sea levels while at other times at the end of glaciations they would have carried much larger volumes of water than is the case today and have been capable of transporting greater sedimentary loads and larger particles than at present. The idealized block diagram

Figure 5.5 A photograph of the face of a sand and gravel pit showing lenses and layers of sand and gravel particles of different sizes. (Aller gravels, Devon.)

and cross-section in Figure 5.6 illustrates the way in which such processes give rise to river terrace deposits. It should perhaps be noted the deposits of this type are typically only a few metres thick but often are of wide geographical extent.

5.4.2 Glacial and periglacial regions

Glacial deposits are formed principally by the processes of physical erosion by moving ice sheets and glaciers and are deposited at the melting ice fronts. They are found in many parts of the world since glaciations have occurred at various times in the geological past. In Britain most of these deposits were formed during the ice ages of the last four million years. The material carried on and in the ice is dumped as ice melts during the recession

Table 5.13 Types of unconsolidated sand and gravel sources

Alluvial deposits	• Typically well sorted in particle size • Relatively clean • Rounded particle shape
Glacial deposits	• Typically variable in shape and particle size distribution • High content of silt and clay • Angular particle shape • Higher proportion of weak and weathered particles • Fluvio-glacial outwash deposits usually the better aggregates
Coastal deposits	• Rarely used, principally for environmental reasons
Marine deposits	• Marine sands and gravels originally laid down as alluvial or fluvio-glacial deposits

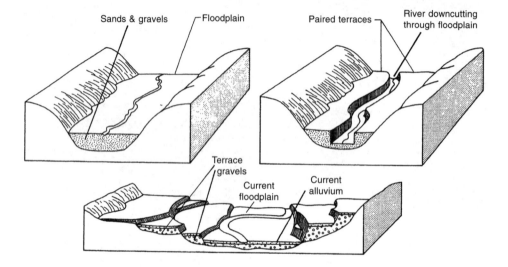

Figure 5.6 The development of river terrace deposits as a river cuts down with time. (After Leet *et al.*, 1982.)

of ice sheets and glaciers in periods of warming between ice ages. The block diagram in Figure 5.7 shows how material is eroded by glaciers and left behind as a glacier recedes during a period of warming. The types of material transported by ice vary considerably, material falling on to and into the ice will tend to be ill-sorted, angular and will cover a wide size range from large boulders downwards to rock flour or clay. This material is referred to as till or **boulder clay**. By contrast, materials carried by meltwater across, through and away from the ice front will have characteristics more similar to those of fluvial or river deposits and are usually referred to as **fluvio-glacial** deposits. As a result of the recent ice ages of the Quaternary much of the British Isles, north of the southern limit of glaciation (Figure 5.4), together with large areas of Russia, Canada and parts of the USA, have suffered the effects of glaciations and retain glacial deposits in valleys and other regions where they have been protected from removal by more recent erosion.

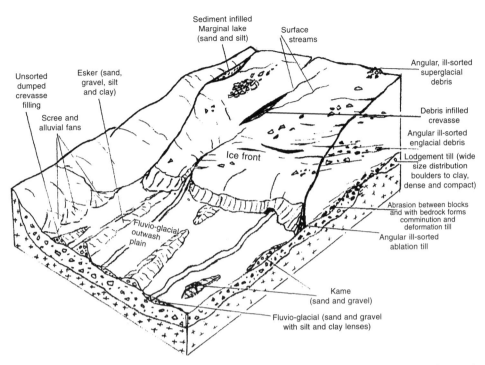

Esker (sand, gravel, silt and clay)

Unsorted dumped crevasse filling

Scree and alluvial fans

Sediment infilled Marginal lake (sand and silt)

Surface streams

Angular, ill-sorted superglacial debris

Debris infilled crevasse

Angular ill-sorted englacial debris

Ice front

Lodgement till (wide size distribution boulders to clay, dense and compact)

Abrasion between blocks and with bedrock forms comminution and deformation till

Angular ill-sorted ablation till

Fluvio-glacial outwash plain

Kame (sand and gravel)

Fluvio-glacial (sand and gravel with silt and clay lenses)

Figure 5.7 An idealized diagram of glacial and periglacial terrain during glacial retreat. (After Fookes *et al.*, 1975 and other sources.)

Calculations indicate that the sea level was lowered in Quaternary times by up to 150 m worldwide during the most extreme of the glaciations. Consequently rivers cut across what is now the shallow seabed around the coasts depositing sand and gravels which are now being extracted by marine dredging. Materials carried from land by icebergs and sea ice during the ice ages have also contributed to these deposits. Current areas where marine dredging is carried out are indicated in Figure 5.4. The depth limits on dredging with current equipment is about 40 m but as technology advances it will soon be feasible to extend this limit to 50 m. However, the land transport distances from suitable quays, the technical difficulties and the numerous environmental constraints all place limits on the economic viability of these resources.

As has been described above, in temperate climatic regions such as that of the British Isles the location of suitable aggregate resources depends either on finding localities where there is suitable rock for quarrying, or on finding fluvial deposits of sand and gravel. In general terms the upland mountainous regions provide possibilities for quarry development and the production of crushed rock aggregate. While sand and gravel resources are usually located in the mature floodplain reaches of river valleys and in shallow coastal waters. These types of topographic expression, and the aggregate resources they might be expected to contain will be broadly similar and applicable in other climatic regions of the Earth, provided the effect of the particular climate on the weathering of rock and the processes of erosion are taken into account.

5.4.3 Hot desert regions

In hot arid climates the processes of erosion are modified in that the breakdown of rock is controlled principally by physical processes such as expansion and contraction due to temperature cycling. The role of water in breaking down a rock is much more limited, though the rare flash floods following rainstorms do provide the energetic mechanism for transporting rock debris from the mountainous regions to be spread as sand and gravel outwash fans at the foot of the mountains, and as alluvial plain silts, sands and gravels on the lower ground. The lack of vegetation and the windy dry conditions favour the development of sand dunes which are often developed on the alluvial plains. The idealized block diagram in Figure 5.8 illustrates these various features, and the kinds of materials typically available in hot desert regions. The high water table level, which is often saline, is also a common feature in the low ground furthest from the mountains, and introduces the additional difficulty of salt contamination of potential aggregate resources.

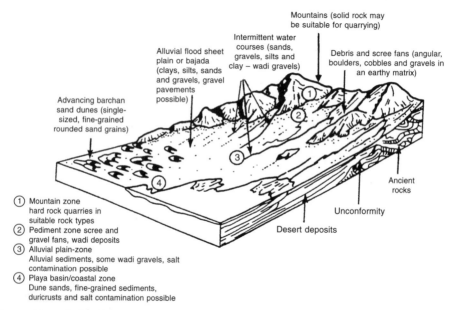

Figure 5.8 An idealized block diagram of a hot desert region showing the landforms and associated materials in the different zones. (After Douglas, 1986.)

The sand dunes may at first sight be thought of as an excellent source of fine aggregate but this is usually not the case because the wind-transported sand particles tend to be very fine-grained, single-sized and poorly graded. Sand resources in the low ground also need to be viewed with caution because salts crystallized within the sand and derived from the saline water table may be present in sufficiently high concentrations to render them unsuitable for concrete aggregate.

Capillary rise of saline ground water can give rise to a salt-cemented layer at the land surface, because the water evaporates leaving the salt as an intergranular cement. Such layers are called **duricrusts**, and may be a metre or so in thickness. If such a duricrust is cemented by carbonate then it may after crushing provide a valuable source of coarse aggregate, but it will be entirely unsuitable as concrete aggregate if the cementing material is halite (NaCl), or gypsum ($CaSO_4 \cdot 2H_2O$) or if the carbonate is contaminated by such minerals.

5.4.4 Tropical hot wet environments

Hot wet tropical climates also introduce differences in the sequence of rock weathering and erosion which in turn is reflected in the types of material produced and the possible sources of aggregates for concrete. The importance of water and elevated temperature on the erosion and weathering in wet tropical conditions is critical in that the zone of rock weathering can develop rapidly, variably and is often very thick. A thickness of 60–100 m of degraded weathered rock is not uncommon and is an important consideration for quarry development. Vegetation also grows quickly adding peat, organic material and humic acids to the soil and ground water so that mature river valleys and alluvial areas are unlikely to yield satisfactory sand and gravel resources. The moist conditions often tend to assist the chemical leaching of rock during the weathering process and lead to dissolution of silica and calcium and leave behind higher concentrations of alumina and iron oxides and hydroxides which may in turn form hard layers within the soil. These layers are called **ferrocrete**, **bauxite** and **alucrete**, depending on their composition. Soluble silica and carbonates may be reprecipitated elsewhere to form **silcrete** or **calcrete** layers. Such hard layers are often exhumed by later erosion to form duricrust layers which can be up to 10 metres thick, they may cap flat-topped hills and form a valuable local source of aggregate. Nevertheless as can be seen in the block diagram in Figure 5.9 the possible localities which may provide sources of concrete aggregates are limited in such terrains.

5.5. Classification of aggregates

The principal requirements for an aggregate to be suitable for concrete are that it will be strong, durable and inert. The individual particles ideally must be of equant shape and be evenly graded from coarse to fine within their particular grading size fraction. Classification of aggregates beyond the broad categories of crushed rock, sand and gravel must be appropriate to its use in the construction industry and have both a scientific and a commercial viability. A petrographic description can be of considerable value in any assessment of the likely performance of a particular rock type, but the geological name is only of importance in that it implies that there will be a set of physical and mechanical properties which are very broadly typical of that particular material. In an early consideration of the practical classification of aggregates the British Standards Institution in 1973 recognized eleven 'Trade Groups' of rocks (BS 812, 1975), each group containing a number of individual rock types classified into their particular group on the basis of the assumed similarities of the physical and mechanical properties they shared. This listing is shown in Table 5.14. Although this concept of grouping rocks together according to their properties appeared to be useful, a number of rock types were difficult to assign to a single group and it omitted a classification for sands and gravels. In general, the variety and range of rock types covered was too wide, with boundaries between groups difficult to define. Also, the use of precise geological names for the groups and names used within groups

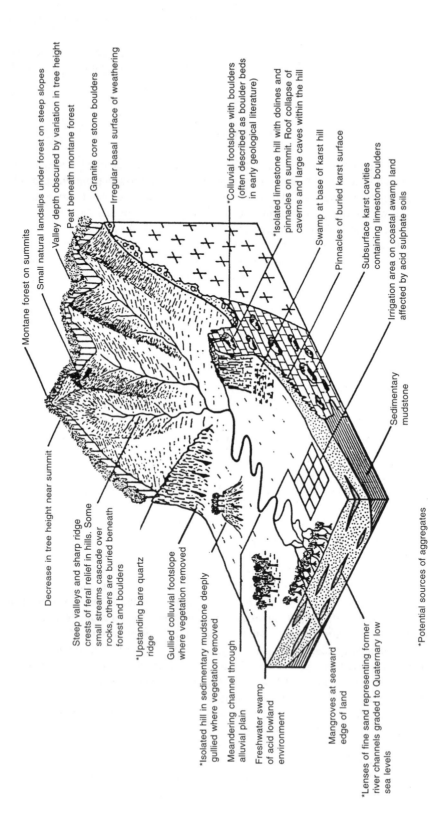

Montane forest on summits

Decrease in tree height near summit

Small natural landslips under forest on steep slopes

Valley depth obscured by variation in tree height

Peat beneath montane forest

Granite core stone boulders

Irregular basal surface of weathering

*Colluvial footslope with boulders (often described as boulder beds in early geological literature)

*Isolated limestone hill with dolines and pinnacles on summit. Roof collapse of caverns and large caves within the hill

Swamp at base of karst hill

Pinnacles of buried karst surface

Subsurface karst cavities containing limestone boulders

Irrigation area on coastal awamp land affected by acid sulphate soils

Sedimentary mudstone

Steep valleys and sharp ridge crests of feral relief in hills. Some small streams cascade over rocks, others are buried beneath forest and boulders

*Upstanding bare quartz ridge

Gullied colluvial footslope where vegetation removed

*Isolated hill in sedimentary mudstone deeply gullied where vegetation removed

Meandering channel through alluvial plain

Freshwater swamp of acid lowland environment

*Lenses of fine sand representing former river channels graded to Quaternary low sea levels

Mangroves at seaward edge of land

*Potential sources of aggregates

Figure 5.9 An idealized diagram of a humid tropical environment. (After Douglas, 1986.)

led to unnecessary contractual disputes, so that the scheme is now omitted from the standards.

The general consensus today is to use the correct geological name for the rock and to take careful account of its mineralogy and state of weathering or alteration. The need to use the correct geological name means that the petrographer in sampling and describing an aggregate for use in concrete must comply with the methods given in appropriate British Standards or their equivalents abroad. In Britain the methodologies are given in BS 812 Parts 102 (1984), 103 (1994), 104 (1994), and 105 (1989) and the European simplified equivalent is BS EN 932–3 (1997). In the USA, ASTM C294–98 and C295–98 (1998) provide a nomenclature and methods for the petrographic description of aggregate. A more specific standard procedure for dealing with the particular issue of alkali–aggregate reaction (AAR) in concrete is given in BS 7943, but this particular topic is considered in greater detail in Volume 2, Chapter 13 of this series. Reviews of special aggregate problems can be found together with reference to the appropriate standards and guidelines in books such as St John *et al.* (1998) and Smith and Collis (2001).

In general terms the accepted recommended approach to the description and classification of any aggregate requires a three-stage description:

1 A description of the aggregate type (sand, gravel, crushed rock)
2 A description of the physical characteristics (particle shape, size, grading)
3 A petrographic description and classification (mineralogy, geological name)

To this simple aggregate classification more detailed petrological information and any test results can be added as is necessary. This approach can be presented simply on a proforma such as is illustrated in Figure 5.10.

5.6 Aggregate quarry assessment

The periodic requirement to inspect a working aggregate quarry, and to sample the aggregates produced so that petrographic examination or other laboratory tests can be undertaken, forms an important part of quality assessment of the aggregate produced.

The most important aspect of any site inspection and sampling programme must be the essential requirement that the observations made, and the samples collected, must be both objective and representative. In carrying out such a programme it is first necessary to hold a briefing meeting with both quarry staff and the inspection team so that the objectives of the inspection are clearly defined and that any other information relevant to the inspection is made available, including information relating to health and safety matters that may affect the inspection procedure. A preliminary tour in the company of quarry staff is invaluable in that attention can be drawn to features of the quarry working that may not be obvious or have become obscured by later operations. This will result in ensuring that no aspect of importance is missed, and consequently time will be saved, while the effectiveness of the detailed inspection is increased. Finding a high vantage point is important in the early stages of the inspection since this will allow the overall geology, geomorphological and geographical setting of the site to be recognized. The past, present and planned future working of the quarry faces need to be determined and its accuracy agreed with the quarry manager. Past, current and future predictions of production must also be estimated and any existing data on earlier test results need to be obtained and

Classification and description of aggregate					
1	**Aggregate type**				
1.1	Crushed rock				
1.2	Gravel		Uncrushed		Land won
			Partly crushed		
1.3	Sand		Crushed		Marine
2	**Physical characteristics**				
2.1	Nominal size				
2.2	Shape				
2.3	Surface texture				
2.4	Colour (sample condition)				
2.5	Presence of fines				
2.6	Presence of coatings				
2.7	Extraneous material				
3	**Petrological classification**				
3.1	Monomictic			Polymictic	
3.2	Petrological name				
3.3	Petrological description				
Visual assessment		Petrographic thin-section		Quantititive analysis	
3.4	Geological age				
4	**Full petrological description**				
5	**Sample ref.**		**6 Certificate of sampling**		
7	**Source**				

Figure 5.10 An example of a simple proforma format for reporting the description and classification of an aggregate, after Smith and Collis, 2001. (Note: 3.1, Monomictic = single rock type, Polymictic = several different rock types.)

evaluated against the current working situation in the quarry. Methods of sampling the quarry faces, the stockpiles of aggregate and the materials on conveyors or at stages in the production process are well established; for example, BS 812, Part 102 (1984) and ASTM D75–87 (1987). The overriding principle behind the procedures selected will be the requirement to obtain a representative sample so that the results of any laboratory investigation or test programme can also be relied on during the period of continued production.

Difficulties will arise in predicting the type and quality of future reserves beyond the current existing faces. The confidence of these predictions will depend partly on the distance the undeveloped material is from existing worked faces and fully characterized material and partly on the potential geological variability of the source. A good geological evaluation of the locality and the type of material, aided by a detailed geological mapping exercise, will in most cases provide the required information to allow an adequately accurate prediction of the nature and quality of the reserves. However, in materials which show wide local variation on a scale of a few metres, such as many unconsolidated sand and gravel deposits, or where reserves need to be 'proved' in detail, a scheme of trenching or pitting will be necessary while in solid rock a pattern of boreholes will be required to sample the variation. In such circumstances the quality of the information obtained is entirely dependent on the care given to logging the pits, or the recovered drill core, and also on the subsequent laboratory examination and testing of the samples obtained.

The technical aspects of developing a new quarry on a 'greenfield' site follows the general pattern of investigation outlined above. In such a case the most important requirement is accurate geological information concerning both the site and the materials. This will be obtained from a desk study, supplemented by detailed engineering geological, geomorphological and hydrological mapping. Once the data have been obtained and evaluated, a programme of sampling using boreholes, pits or trenches as appropriate must be designed for the site together with a detailed plan for the laboratory examination and testing of the samples obtained.

5.7 Deleterious materials in aggregates

The natural rock and gravel sources of aggregates may contain components which are potentially deleterious when the aggregate is used in concrete and mortar. These deleterious materials may be distributed throughout the entire rock or deposit, be confined to only a part of it, or alternatively be present in a particular localized feature within the source rock or deposit. A geological evaluation of the source combined with a thorough petrological investigation and appropriate testing will minimize the risk of deleterious materials being incorporated into the concrete or mortar. A listing of the most common potentially deleterious materials has been assembled by Sims and Brown (1998) and is presented in Table 5.14.

In addition to deleterious materials being incorporated into the aggregate from its source, other components may find their way into an aggregate stockpile as contaminants. These may be such materials as metal, wood and plastic fragments introduced during the extraction and processing of the aggregate or as chemical solutions such as salts derived from the capillary rise of saline ground waters into the base of aggregate stockpiles, or from salt spray or contamination from adjacent chemical storage facilities. Again, in some rare cases the processed aggregate degrades after it has been processed either in the stockpile, or in the concrete made with it. This indicates that the constituent minerals in the aggregate particles themselves are unstable when exposed to air and water from the atmosphere.

The identification and assessment of the concentration of any of these deleterious materials in an aggregate can be achieved by a combination of petrological, physical and chemical techniques. The results of such a study may indicate that further careful investigation of the source of the aggregate, its processing or its storage will be necessary in order to identify and isolate or remove the deleterious component.

Table 5.14 Some potentially deleterious constituents found in aggregates

Potentially deleterious constituent	Possible adverse effect in concrete[1]				
	i	ii	iii	iv	v
Clay coatings on aggregate particles	–	√	–	–	–
Clay lumps and altered rock particles	–	–	✓	✓	✓✓
Absorptive and microporous particles	–	–	✓	✓	✓✓
Coal and lightweight particles	–	–	–	–	✓✓
Weak or soft particles and coatings	–	✓✓	✓	✓	✓✓
Organic matter	✓✓	–	✓	–	–
Mica	–	–	✓✓	–	✓
Chlorides[2]	–	–	–	✓	–
Sulfates	–	–	–	✓✓	✓
Pyrite (iron disulfide)	–	–	–	✓✓	✓✓
Soluble lead, zinc or cadmium	✓✓	–	–	–	–
Alkali-reactive constituents	–	✓	–	✓✓	–
Releasable alkalis	–	–	–	✓✓	–

1   (i)   Chemical interference with the setting of concrete
   (ii)   Physical prevention of good bond between the aggregate and the cement paste
   (iii)   Modification of the properties of the fresh concrete to the detriment of the durability and strength of the hardened material
   (iv)   Interaction between the cement paste and the aggregate which continues after hardening, sometimes causing expansion and cracking of the concrete
   (v)   Weakness and poor durability of the aggregate particles themselves
2   The main problem with chlorides in concrete is associated with the corrosion of embedded steel.
✓✓ = main effect ✓ = addition effect

5.7.1 Interference with the setting of the concrete

Certain chemical contaminants such as soluble chlorides and sulphates, organic compounds such as mono- and polysaccharides, humic acid and lignins, and also soluble salts of lead, zinc, cadmium and tin, can retard the setting of cement, and hence affect the final strength and durability of concrete. Even in low concentrations their effects are noticeable and, in extreme cases, they can prevent the proper setting of the cement binder so that the concrete is friable or crumbly and has little or no strength. In the case of some of these contaminants, notably sulphates and chlorides, they remain a potential problem and threat to durability long after the concrete has set.

Chemical analysis is clearly a means of identifying such contaminants but petrographic techniques such as the selective staining of sulphate minerals, the X-ray diffraction identification of deleterious crystalline materials, and infra-red spectroscopy or the use of scanning electron microscopy plus a microanalyser for the identification of chemical phases are all very helpful tools in the identification of small amounts of deleterious chemical components.

The upper limits for soluble chloride specified in BS 882 (1992) range from 0.01 per cent to 0.05 per cent depending on the type of concrete. Problems with chloride contamination of aggregates are rare in the UK since aggregates are usually washed, but in some climatic regions, notably the Middle East, chloride contamination is not uncommon. Chloride reduces the protection of reinforcing steel from corrosion (passivation) and leads to rust forming on the surface of the steel. The rust occupies a much greater volume than the original steel thus disrupting and cracking the concrete as well as weakening the all important concrete–steel bond and the steel itself. Carefully measured amounts of gypsum (hydrated calcium sulphate) are normally added to Portland cement to control its setting characteristics, but additional amounts derived from the aggregate could produce deleterious expansion and disruption of the concrete through internal sulphate attack. Such damage can be readily identified in a petrographic thin-section (Figures 5.11–5.13). Soluble sulphate is rarely a problem with UK aggregates. However, sulphates such as gypsum may form coatings as is illustrated in Figure 5.14, or may be present as discrete particles in aggregates and may be a serious problem with aggregate sources in some climatic regions, especially hot dry deserts.

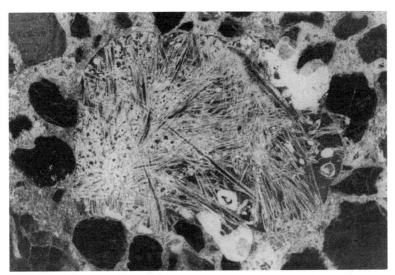

Figure 5.11 A photomicrograph of a petrographic thin-section showing needle-shaped crystals of ettringite in a void, the result of conventional sulphate attack.

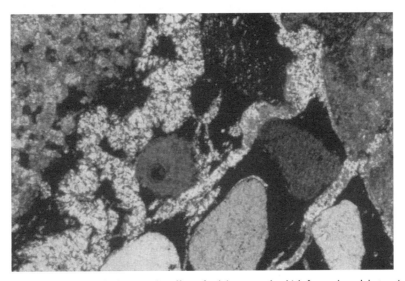

Figure 5.12 A photomicrograph showing the effect of sulphate attack which forms the sulphate mineral, thaumasite, which infills the irregular cracks.

Figure 5.13 A severe example of the thaumasite type of sulphate attack in a hand specimen of concrete.

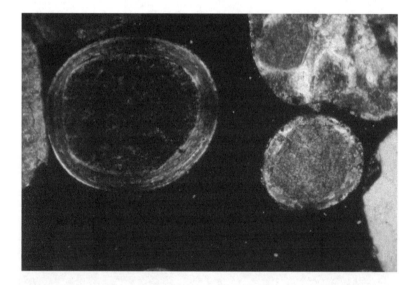

Figure 5.14 A photomicrograph showing rounded fine-aggregate sand grains with a coating of crystalline gypsum.

Metal cations such as lead and zinc are known to interfere with the setting of cement. However, problems are rarely encountered in practice and only occur if aggregates have been contaminated with soluble metal compounds, or where recycled metaliferous mining wastes have been used as aggregate.

5.7.2 Modification to the strength and durability characteristics of a concrete

Although small amounts of dust are helpful in concrete, improving cohesiveness and reducing bleeding, excessive amounts may cause the aggregate particles to become coated with dust so that the necessary bonding between the particle and the cement matrix is not fully established. As a consequence the concrete will have poor durability and strength characteristics from early ages. Coatings of clay or other weak material can have similar effects.

Interaction between aggregate particles in a concrete or mortar and constituents from the cement binder can also lead to longer-term durability problems. The best known of these are the alkali-aggregate reactions between certain types of siliceous or carbonate aggregates and the alkalis normally derived from the cement. The alkali silica reaction involves aggregates which are composed of (or contain) non-crystalline or poorly crystalline silica or siliceous glass which can react with the alkali hydroxides derived from the cement. In time, the reaction product which is an expansive hydroscopic gel, can exert sufficient pressure to crack and disrupt the concrete. Certain impure carbonate aggregates will also react with alkalis in the pore fluid (referred to as the alkali carbonate reaction) and can also cause expansion and disruption of the concrete. Both these reactions between alkali and aggregate in concrete are dealt with in Volume 2, Chapter 13 of this series.

The microporosity of aggregate particles may have a significant effect on the strength, durability and freeze–thaw resistance of a concrete. However, it is difficult to generalize since some porous aggregate materials appear to have beneficial effects when used in a concrete. The nature of the pore structure and pore size distribution, particularly the microporosity, appears to be more critical than the overall porosity of the aggregate in relation to its resistance to freeze–thaw in concrete, so that each aggregate source needs to be assessed individually. Some flints which occur in southern England have a white microporous flint outer layer or cortex which may be present as discrete particles or surface coatings in an aggregate and these can cause spalling and pop-outs from concrete surfaces under freeze–thaw conditions as is illustrated by Figures 5.15–5.19.

5.7.3 Unsound aggregate particles in concrete

A wide range of aggregate materials contain unsound particles, present either as natural components in the aggregate source material or as contaminant particles from an extraneous source. They can produce problems when incorporated into a concrete which range from surface cosmetic disfigurement to reduction in the compressive strength, or a reduction in durability.

One of the commonest unsound particle contaminants familiar in the gravel aggregates of southern Britain is pyrite (iron disulphide). This mineral can oxidize to brown iron hydroxides in normal environmental weathering conditions if close to the surface of a concrete. The ease with which this oxidation takes place depends on the purity and the detailed crystal structure of the mineral, but can lead to development of brown staining on the concrete surfaces illustrated in Figure 5.20. Although such stains do not normally affect the structural integrity of dense concrete, the unsightly stain is usually much larger than the reacting mineral particle.

Figure 5.15 Examples of flint and chalk (extreme right) aggregate particles showing the white microporous cortex on the flints (partly broken away on the particle second from left).

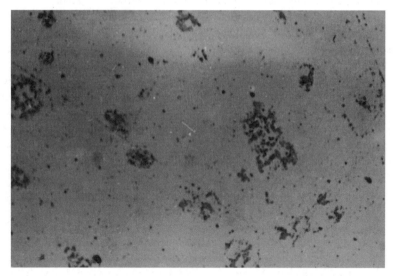

Figure 5.16 A scanning electron microscope image of dense flint spares are the black areas.

However, some Cornish tin mining wastes contain pyrite crystals, or 'mundic' particles, (mundic is the old Cornish word for pyrite). This material is used as low-grade aggregate for the production of concrete blocks and the pyrite does contribute to the deterioration of concrete, although moisture movement in the clay and the secondary micas which are present in the slaty and phyllitic waste is also involved. Examples of the type of deterioration are shown in Figures 5.21 and 5.22. Guidance notes concerning this particular 'mundic' problem have been published by the Royal Institution of Chartered Surveyors (1997).

Aggregate particles partly composed of clay and certain geologically weathered rocks

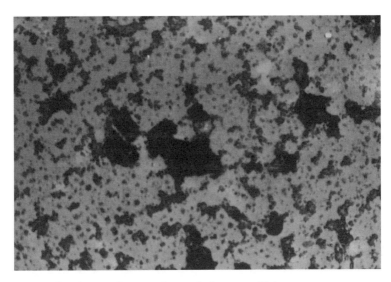

Figure 5.17 A scanning electron microscope image of microporous flint.

Figure 5.18 Surface pop-outs caused by microporous flint aggregate in the concrete.

tend to be moisture sensitive, that is, they have the capacity to absorb or lose water from their structure depending on the environmental conditions. This produces a consequent expansion or shrinkage of the particle leading to the breakdown of the aggregate/cement bond so that the concrete becomes susceptible to damage by frost and thermal cycling. This problem of aggregate shrinkage was recognized in Scotland in the 1950s (Edwards, 1970). Attempts were made to list the petrological rock types susceptible to shrinkage, but this proved to be too broad and unreliable. The petrological evaluation of individual aggregate materials taking account of factors such as weathering together with testing (for example, BS 812, Part 120, (1990) or ASTM C157-99 (1999)) are necessary to establish the stability of an aggregate for use in concrete.

Figure 5.19 Scaling of a concrete surface owing to freeze–thaw, exacerbated by de-icing salts.

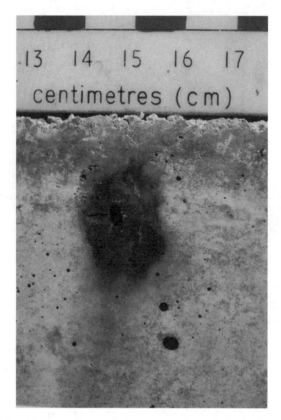

Figure 5.20 Iron staining due to a pyrite particle present near the surface of a concrete.

Figure 5.21 An example of a degraded 'mundic' block.

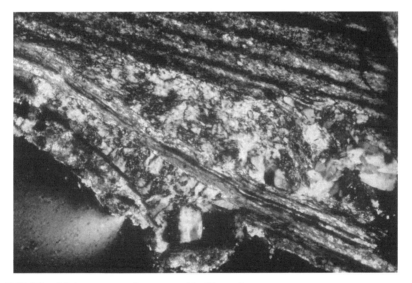

Figure 5.22 'Mundic' decay as seen in petrographic thin-section.

Chalk particles present in some gravel aggregate sources (see Figure 5.15) can also reduce the quality of a concrete. Chalk aggregate particles, like clay lumps and other soft materials, are weak and can reduce the compressive strength and durability of the concrete. However, a fraction of the chalk appears to be broken up and becomes involved in the hydration and setting of the cement matrix with uncertain consequences. Thus only a proportion of the original chalk particles remain as true aggregate, nevertheless a small percentage of chalk in the aggregate is sufficient to reduce the strength of a concrete significantly.

Varying amounts of coal and lignite particles are found in some gravel aggregates and also in some mine wastes. They tend to form weak and porous particles in the aggregate, and may also form unsightly tarry stains on concrete surfaces, or be the cause of surface pitting. An example of coal aggregate particles in petrographic thin section is shown in Figure 5.23.

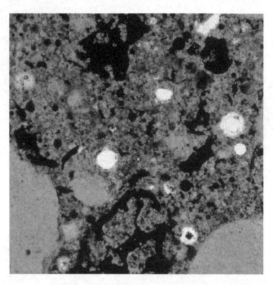

Figure 5.23 A petrographic thin-section photomicrograph showing porous coal fragments in a concrete.

The micas, biotite and muscovite, are common flaky constituents of many igneous and metamorphic rocks. The processes of erosion acting on such rocks tend to produce sands which contain varying proportions of discrete mica particles, particularly the colourless muscovite mica which is the more stable, and does not readily break down. Mica as a constituent of a sand fine aggregate usually increases the water demand of a concrete and because of flaky shape and surface texture tends to reduce the cohesiveness of the mix. Small percentages of mica in the fine aggregate can reduce the final compressive strength of a concrete significantly. In one example from south-west England strength was reduced by 5 per cent for 1 per cent by weight of discrete mica particles in the total aggregate. However, such reductions in strength can be offset by using higher cement contents or admixtures.

Many sea-dredged and some land-based gravel resources contain carbonate shell fragments. Although these fragments are usually mechanically strong complete shells may be hollow and other fragments flaky. The nature of these fragments may reduce the workability of a concrete mix and hollow shells may reduce the strength of the concrete. However, unless the proportion of shell in the aggregate is very high the effects of shell on the workability and strength of concrete are minimal.

References

American Society for Testing and Materials (1988) Standard descriptive nomenclature of constituents of natural mineral aggregates, ASTM C294–98.

American Society for Testing and Materials (1999) Testing and Materials (1999) Test for length change of hardened cement, concrete and mortar, ASTM C157–99.

American Society for Testing and Materials (1998) Standard practice for petrographic examination of aggregates for concrete, ASTM C295–98.

American Society for Testing and Materials (1992) Standard practice for sampling aggregates. (Re-approved 1992), ASTM D75–87.

British Standards Institution (1975) Methods for sampling and testing of mineral aggregates sands and fillers, BS 812. Parts 1–3.

British Standards Institution (1984) Methods for sampling, BS 812, Part 102.

British Standards Institution (1985) Methods for determination of particle size distribution, BS 812, Part 103.

British Standards Institution (1989) Methods for determination of particle shape, BS 812, Part 105.

British Standards Institution (1989) Method for testing and classifying drying shrinkage of aggregates in concrete, BS 812, Part 120.

British Standards Institution (1990) Methods for determination of drying shrinkage, BS 812, Part 120.

British Standards Institution (1990) Specification for lightweight aggregates for masonry units and structural concrete, BS 3797.

British Standards Institution (1992) Specification for aggregates from natural sources for concrete, BS 882.

British Standards Institution (1994) Procedure for qualitative and quantitative petrographic examination of aggregates, BS 812, Part 104.

British Standards Institution (1997) Tests for general properties of aggregates, Part 3: procedural terminology for simplified petrogaphic description, BS EN 932–3.

British Standards Institution (2002) Aggregates for concrete, BS EN 1260.

Carmichael, R.S. (1982) *Handbook of Physical Properties of Rocks*, Volume 1. CRC Press, Boca Raton, FLA, Table 17, p. 34.

Concrete Society (1999) Alkali silica reaction. Minimising the risk of damage to concrete. Guidance notes and model specification clauses. Report of a Concrete Society Working Party. Concrete Society Technical Report No. 30, 3rd edn.

Douglas, I. (1986) Hot wetlands. In Fookes, P.G. and Vaughn, P.R. (eds), *A Handbook of Engineering Geomorphology*, Blackie, Glasgow (Surrey University Press).

Eagar, R.M.C. and Nudds, J.R. (2001) *The Geological Column*, 8th revised edn. Printguide Ltd, Manchester, Panel 1.

Edwards, A.G. (1970) Shrinkable aggregates. Building Research Station Scottish Laboratory, East Kilbride, SL **1**, (1) 1–70.

Folk, R.L. (1959) Practical petrographic classification of limestones. *AAPG Bulletin*, **43**, 1–38.

Fookes, P.G. (1989) Civil engineering practice. In Blake, L.S. (ed.), *Civil Engineer's Reference Book*, 4th edn. Chapter 8, p. 8/9.

Fookes, P.G. and Higginbottom, I.E. (1975) The classification and description of near-shore carbonate sediments for engineering purposes. *Géotechnique*, **25**: 406–411.

Fookes, P.G., Gordon, D.L. and Higginbottom, I.E. (1975) Glacial landforms, their deposits and engineering characteristics, in the engineering behaviour of glacial materials. *Proceedings of a symposium of Midland Soil Mechanics and Foundation Engineering Society*, reprinted by *Geoabstracts*, Norwich, 1978.

Fookes, P.G. and Higginbottom, I.E. (1980) Some problems of construction aggregates in desert areas, with particular reference to the Arabian Peninsula. I – Occurrence and special characteristics. II – Investigation, production and quality control, *Proc. of the Institution of Civil Engineers*, Part 1, **68**, Feb. pp. 39–90.

Fookes, P.G., Dearman, W.R. and Franklin, J.A. (1971) Some engineering aspects of rock weathering with field examples from Dartmoor and elsewhere. *QJEG*, **4**, 139–185.

Fox, R.A. (2002) Non-energy mineral resources. In *Maritime World 2025, Future Challenges and*

Opportunities, The Greenwich Forum conference proceedings 3–5, April 2002, available on CD-ROM.

Geological Society Engineering Group (1995) The description and classification of weathered rocks for engineering purposes. *QJEG*, **28**, 207–242.

Kirkaldy, J.F. (1954) *General Principles of Geology*. Hutchinson, London, pp. 1–327.

Leet, L.D., Judson, S. and Kaufman, M.E. (1982) Physical Geology, 6th edn, Prentice Hall, Englewood Cliffs, N.J, pp. 284–285.

Mclean, A.C. and Gribble, C.D. (1990) *Geology for Civil Engineers*, 2nd, edn. Unwin Hyman, London.

Royal Institution of Chartered Surveyors (1997) The 'Mundic' problem – a guidance note – recommended sampling, examination and classification procedure for suspect concrete building materials in Cornwall and parts of Devon. 2nd edn (Chairman) Stimson, C.C., pp. 1–75.

Smith, M.R. (ed.) (1999) *Stone: building stone, rock fill and armourstone in construction*. Geological Society Engineering Geology Special Publication No. 16, London, pp. 1–478.

Smith, M.R. and Collis, L. (eds) (1993) *Aggregates, sand, gravel and crushed rock aggregates for construction purposes*, 2nd edn. Geological Society Special Publication No. 9, London.

Smith, M.R. and Collis, L. (eds.) (2001) *Aggregates, sand, gravel and crushed rock aggregates for construction purposes*, 3rd edn. Geological Society Special Publication No. 9, London.

St John, D.A., Poole, A.B. and Sims, I. (1998) *Concrete Petrography. A handbook of investigative techniques*. Arnold, London.

Sims, I. and Brown, B.V. (1998) Concrete aggregates. In Hewlett, P.C. (ed.), *Lea's Chemistry of Cement and Concrete*, 4th edn. Arnold, London, Chapter 16.

Winchester, S. (2001) *The Map that Changed the World*. Viking, Penguin Books, London.

6
___

Aggregate prospecting and processing

Mark Murrin-Earp

6.1 Aims and objectives

This chapter describes the processes involved in the winning of aggregates and their processing, looking at the different methods employed in doing so and the machinery and equipment used. Predominantly the areas covered will be:

- Extraction and processing of sand and gravel
- Extraction and processing of limestone.

It must be realized that generic examples will be used for the layout of different extractive processes, using typical machinery and equipment and that naturally occurring sources of raw materials vary. Therefore, when approaching a specific deposit, it may be necessary to employ modifications to the types, setup and usage of such equipment. The sections will be divided into:

- Extraction.
- Processing and machinery and equipment.

6.2 Introduction

As long as society demands that buildings and roads be constructed, then naturally aggregates will be extracted to produce the materials required. The aggregate extraction industry has grown to keep pace with demands for aggregates for use in construction materials and since the Second World War the volume of materials extracted has increased steadily, with only brief deferments for periods of global and national recession. Today, modern sophisticated machinery and equipment are used to produce high volumes of materials.

The characteristics and properties of the aggregate can go a long way to dictate the final performance of the material. Obviously this will mean differences in the qualities of aggregate sources and will influence the selection of the aggregate source for its final use.

Aggregate quality, location and mineral type will also determine the mode of extraction and the processing that it will subsequently receive post-extraction.

Aggregate is a mass-produced commodity and therefore must be a cost-effective raw material to the marketplace. However, because mass production techniques are employed to produce cost-effective construction material sources, this does not mean that quality issues can be overlooked. High levels of capital investment by companies in sophisticated static and mobile machinery and equipment results in high quality and consistent products.

6.3 Extraction and processing of sand and gravel

First, in the case of wet and dry extraction of sand and gravel consideration must be given to the removal of any overburden from the deposit. It is important to realize the value of accurate site investigation prior to the commitment to extract.

Other concerns in this area will centre around any costs related to the removal of any overburden and the implications of any further sterilization of deposit by the removal and subsequent stocking of this material, along with its use in any future restoration of the finished site. This last factor of restoration is again a consideration in terms of the final costs of completion of the site.

The extraction of land-won sand and gravel consumes a large amount of land due to the relatively shallow nature of most deposits of these types. Most land-based deposits are usually only, at worst, semi-bound (relates to the degree of cementation of the deposit) or typically unbound. It is common to strip overburden in a phased manner, in order to minimize the risk of double-handling of material and to remain cost effective. The use of good estate management practice is essential if this is to be kept to a minimum.

The three main methods of extracting sand and gravel are as follows:

- Wet
- Dry
- Marine

Wet extraction refers to a deposit below water table level, with the material to be extracted in a constantly 'wet state'. This mode of extraction requires equipment that is largely exclusive to this deposit in the case of suction dredging and floating cranes or grabs. Draglines may be also used, which in turn is a commonly used method to extract 'dry deposits'.

The use of tracked or walking draglines has certain disadvantages in terms of the depths of extraction that can be worked, economically with depths of between 5 and 10 metres. Loss from the bucket is also a factor. Generally we must consider the amount of additional conveyor line or dump trucks, which may prove problematic when workings are of any distance from the site allocated from the processing plant.

If choosing standard or multiple-axle articulated dump trucks to deliver won material from a dragline, care must also be exercised so that the correct selection of capacity is achieved. This allows the primary extraction device to have an equal amount of bucket passes from the face to the dumper so as not cause under- or overloading of the vehicle. This will result in efficient material transfer to the processing plant. Multiple 'shifts' of vehicles may be required to turn around high volumes of won material for processing. Alternatively the material can be temporarily stocked to allow some free draining to take place. Generally the use of vehicles to transport material any distance is inefficient and, wherever possible, conveyor transport should be used.

One of the alternatives to the above is using a suction dredger. Sites of this nature where this system is to be employed as the means of extraction must be looked at with care, as they require years of extraction to make the initial investment worthwhile.

Finally the use of floating cranes with a fitted grab or a ladder dredger (inclined conveyor system which has buckets fitted to allow the material to be scraped from the bed of the submerged deposit) in this environment is also suitable, where a barge-mounted crane with a grab-type bucket is fitted. Usually to ensure that the operation is effective it is necessary to have some form of surge hopper fitted to the barge, which in turn is connected to land via a conveyor and pontoon arrangement. This allows the barge to move around different areas of the submerged deposit easily once an area has been worked to the desired depth. Again with this type of deposit, a relatively shallow extraction depth can be achieved. So once again a commitment to a large area of extraction is required. Restoration planning on this type of site will usually mean a lake as obviously no overburden stripping is required and therefore there will be no surplus materials to effect a fill. The won aggregate can be pumped ashore or fed into barges, which in turn are then landed to deliver the material for processing.

Generally these types of extraction are on a large scale, with higher volumes of material involved. 'Who wants to buy a second-hand land-locked boat?' Anon., 1997).

Dry 'pit' extraction refers to deposits that are either naturally dry or which are extracted from land that has been drained to facilitate dry extraction. Again in this environment either dragline or face shovel extractive means can be employed, depending on the volume required. Limitations for draglines again apply, so to be effective, extraction deposits of 5 m in thickness or greater are required. These are efficient and are therefore widely used.

Face shovels are another option, and prove efficient in unbound or semi-bound extractive environments.

Variations on hydraulic machinery such as the use of back-acting or backhoe excavators also have advantages and disadvantages within themselves, in that they are capable of digging semi-bound dry deposits. The main disadvantage compared to a dragline is the depth of formation to which they can dig. However they are extremely useful for forming bunds and barriers on-site, as well as assisting in the reinstatement of finished areas.

The size of the pit will depend on the number of hydraulic excavators used. Typically in a medium-sized pit producing 100 000 tonnes or more two faces can be worked. Extracted material is then transported to the primary processing plant or, depending on the site layout, a surge pile or hopper.

Moving the raised materials to processing again tends to favour high-speed conveyors, which are fixed, and they in turn are fed from branch conveyors that allow for movement and re-assembly to the live face working area of the quarry. The most important factor is to minimize the use of vehicular transportation, which in turn reduces operating costs such as wear and tear on expensive capital items of plant. Again, as with wet extraction, we consider the ratio of the hydraulic loading device to the size of dumper used. For instance, if a long-reach back-acting unit is used, in order to reduce the stresses on the arm, a smaller bucket is fitted. Therefore it may require several passes in order to fill a larger dump truck. In this case one needs to carefully design the extraction to maximize the use of equipment.

When opening a new dry or wet pit, it may also be necessary to use mobile and not static plant and equipment in order to begin the works. Such equipment can be installed periodically to boost production of a particular material and to assist the existing static plant. This may be done by engaging a suitable contractor, or by hiring the equipment from a plant hire company. These measures are to be viewed as temporary and may require some addressing regarding planning permission to have the site for longer periods of time.

A wide variety of equipment is available, and if the extraction is for a limited time, this may well prove an option, particularly if the pit is producing a small range of different types of material in high volume. Considerations in terms of locating this plant on-site are required, again as previously stated to maximize output and efficiency.

Marine extraction is by suction dredger and takes place in river estuary environments and continental shelves and as a means of extraction this method has grown in popularity.

Investment is initially high, as construction of a specialist ship is required as well as a suitable wharf for landing of the raised aggregates.

The wharves are also ideally located close to the demand for the materials and within reach of the source, to effect a collect and discharge cycle. This allows the won aggregate to drain and be processed while the dredger returns to sea to reload. Transport by ship is also increasing in popularity as the economies of scale come into play, with high volumes being transported.

There are two main modes of operation for dredging fleets to win aggregates:

- Dredging at anchor with a forward leading suction pipe
- Trail dredging with the ship underway.

Dredging at anchor implies that the ship is anchored and the suction pipe is lowered to the seabed in forward mode. This forms a 'sea pit' from the initial lowering and commecement of the operation. Once this pit has been formed and because the aggregate being extracted exhibits fluid properties at this stage, the pit on the seabed is continually replenished from the aggregate immediately surrounding the excavation. Periodically the ship will need to move forward to form another pit and the process can then recommence.

The main advantage of this is that it allows for deeper extraction depths to be achieved, but it has environmental advantages and disadvantages as well.

Unlike dredging at anchor, *trail dredging* involves the ship being underway and, as the name suggests, a trailing suction pipe is lowered to the seabed. As the ship sails, aggregates are pumped or sucked from the deposit by the use of hydraulic pumps. These pumps create a head of vacuum that draws up the aggregate. Depths are somewhat limited with type of operation, with operating depth between 25 and 45 m.

A certain amount of processing, by way of washing of some unwanted fines from the raised material, can take place at sea. Generally at sea processing is not permitted.

Extraction licenses may also be required to remove materials from certain environments, particularly if within close proximity to fishing grounds. However, river estuary dredging for navigation purposes can have significant benefits. The main drawback is the potential quality and overall suitability of such materials.

6.4 Processing

The level of processing affecting sand and gravel deposits will often depend on how that material was originally won. That aside, there are similar attributes to the eventual processing that the aggregate will receive.

The purpose of processing aggregates is so they can be used in other products, and the type and level of processing they receive, ranging from simple screening to screening, crushing and washing, has an influence on the finished quality of the aggregate and its overall compliance with relevant standards.

There are three main areas of processing with either of the three means of winning sand and gravel:

- Crushing
- Screening
- Washing

Crushing will largely depend on the percentage of coarse aggregate in the deposit, i.e. the amount of material deemed to be retained on a 5.0 mm test sieve.

Some deposits, usually dry pit river terrace deposits such as the Thames Valley, have up to 30%-plus gravel in the deposit. The location of the pit by geological site investigation is important, as knowing the ratio of sand to gravel in the deposit validates the choices to make regarding the selection of processing plant for the site. These ratios are determined by digging trial pits on the site before any quarrying takes place. Subsequent samples from these trial pits are graded in the laboratory so that an overall estimation can be made regarding the levels of coarse and fine aggregates in the deposit. If there were an over-abundance of coarse aggregate in the deposit, for instance, this would probably indicate the unsuitability of extraction taking place from this location.

Also the level of processing of the materials raised from the deposit can depend on the degree of the bound state of the aggregates.

The most effective method of moving aggregates around the site is undoubtedly by mechanized conveyors. High-speed conveyors can move large volumes of materials in short spaces of time, and the selection of type and size of the system will depend on its location within the process. Material from the primary crushing unit will therefore obviously require a wider, high-volume conveyor than, say, one moving material in between washing or screening operations. A conveyor is essentially a straight, continuous section of vulcanized rubber layers bound onto nylon webbing and is extremely durable. This belt is tensioned and held in place by the head rollers and kept lifted by smaller support rollers. Adjustment of the head roller tension causes the belt on the roller to skew in different directions, so this makes it possible to keep the unit well maintained and running true. Failure to ensure correct belt alignment will cause premature wearing of the belt and additional maintenance costs. If moving materials on an incline the belt can have V slots vulcanized to the belt to allow the material being moved to grip the belt. Head and tail main drums can also be fitted with a scraping device to prevent the build-up of detritus on the belt, in the same way that head and tail rollers are required to be kept clean to avoid uneven wear and tear on the unit.

Aggregate is moved in different directions by the use of transfer points, which are effectively smaller hopper-like enclosures that prevent material spillage around the point. Under health and safety regulations there will need to be guarding of the moving parts and a means of tripping the conveyor in the event of an emergency stop being required.

Crusher selection and types are, again, largely dependent on the ratio of coarse and fine aggregates present in the deposit and the distribution of particle sizes within it.

It is also possible to carry out some pre-screening of the as-raised aggregates in order to remove undesired elements such as large cobbles, wood, lignite and clay nodules. This is specialist processing that requires particular specific equipment relative to the contaminant in the deposit and will be covered by a later section.

Once the primary considerations for the type of deposit in question have been considered then the type of crusher for the reduction in size of the coarse aggregates in the deposit can be selected.

There are three main types of crusher to select from, with some variations on their individual themes:

- Jaw crusher
- Impact crusher
- Cone crusher

A jaw crusher functions by having a fixed jaw and a moving jaw, the material exiting the crusher at the desired size by setting the gap between the two jaws. The exit gap is set by means of adjusting a bolt and spring assembly at the unit's base, hence allowing relatively fine adjustment to the exit gap and the size of the material exiting. This is known as the reduction ratio.

The crushing motion is by means of a flywheel that causes the moving jaw to swing. The moving jaw can have either single or double toggles fitted which give it additional movement in either two or three planes. This gives the moving jaw greater flexibility and is less prone to jamming with oversized feed. The crushing motion is that of a powerful compressive force caused by the reciprocal motion of an eccentric shaft.

The jaw crusher is almost a universal design and is found and indeed is suitable in many locations. It is often used as the primary means of crushing in many pits and quarries and is best choke-fed. Careful selection of the primary feed material for this type of crusher is vital, as they are prone to jamming with oversize. It may be necessary to use a pre-crushing screen to avoid this. Some crushers of this type can be fitted with pneumatic breakers, in order to unblock any jams that occur during the production cycle. The wear parts are usually of highly wear resistant manganese steel.

The main advantage of this type of crusher is that they are hard-wearing (requiring breaker plates on the jaws to be periodically replaced: this will depend on the abrasive nature of the material being crushed, and the general rule of thumb on this is the *lower* the aggregate abrasion value, the more wear and tear will be on the crusher's wear parts). They are easy to maintain and are a cost-effective medium volume solution to aggregate production.

The disadvantages are that they cannot achieve very high volumes, they produce average quality shaped aggregate, which tends to be elongated, and they are prone to jamming, which causes downtime in the production day.

They can be fed directly from the primary separation of coarse and fine aggregate fractions screening process.

With regard to maintenance, it is usually prudent to carry a spare set of jaws in store as part of any successful extraction business's planned preventative maintenance scheme. It is also possible to effect temporary repair to the wear parts of the jaws by applying strips of weld vertically to the wear part plates, but only as a temporary measure. These types of crusher are also relatively cost-effective to run in terms of power demand once they are up to speed. The greatest power demand is during start-up, when the motors will have their highest current draw. The flywheel has an eccentric drive from the standard V-belts that come from the motor; the amount of belts depends upon the size of the crusher.

Sometimes called 'cone' crushers, gyratory crushers give a medium output in terms of processing and, unlike jaw crushers, provide an excellent means of shaping aggregate due to their action. They consist of a lined external cone and an external cone set on an eccentric shaft bearing to produce the crushing motion. However, the shaft does not rotate but gyrates instead. This makes good compatability for reduction in size of hard and abrasive rock types. Only certain ratios of reduction can be applied, with the reduction of larger to smaller sizes (40 mm and 28 mm to 14 mm and 10 mm) an ideal choice.

They are best located as part of a secondary crushing arrangement, with a means of diverting other sizes via a chute mechanism to allow diversion from the screening process. They are rarely used in primary mode like other types of crusher due to the relatively low volumes produced; the entrance aperture at the top of the crusher is also the limiting factor as to the size that can be fed.

In terms of maintenance, the cone and lined wear parts consist of manganese steel. If not set to gyrate at an optimum amount of cycles (usually between 100 and 200 revolutions per minute) and feeds are not correctly balanced, this will cause uneven wear and poor performance. Once set to run correctly they can run for long periods without adjustment.

Adjustment of the exit gap for the finished aggregate to be fed to the screening process can be via either mechanical or, with more modern units, hydraulic. This involves closing down the exit aperture by either of the above means and allowing the aggregate to be retained longer in the crushing chamber. The more accurate way to control the exit shape of material is, again like the jaw crusher, to steadily choke feed it. Overfeeding will mean spilt and lost material, while underfeeding will cause low output and poor shape. In general, all new gyratory crushers have hydraulic gap-setting devices, which is again less labour-intensive.

Impact crushers can operate in two planes, horizontally and vertically. They are high-volume devices, often found located as the primary crushing unit. The crushing action is by a rotor, with 'swing-out' blow-bars fitted. When fed they 'throw' the primary feed material against a series of breaker plates. The exit gaps, which can be a series of bars, prevent the material exiting before it has reached the desired size. An impact crusher is, like the gyratory crusher, conducive to producing good shaped almost cubical aggregates.

The breaker plates, speed of the rotor and blow-bar type are all adjustable, which makes this a versatile device. Adjustment of the breaker plates is by hydraulics, which move the plates further or nearer the rotor. The nearer the plates are to the rotor, the more energy is dissipated into the crushing action. This will mean a smaller product if desired or, conversely, a longer time to produce, with longer spent in the crushing chamber.

Horizontal impact crushers have one main disadvantage in that they are prone to wear. Changing the breaker plates can either be by turning round the plates, or when they have been used, by new plates. Because there are so many plates employed, this takes time and, ultimately, will mean total replacement of the rotor itself due to excessive wear and tear of extremely abrasive rocks. However, in moderately abrasive environments, they demonstrate good wear and value for producing volume aggregate from the primary feed to the screening process. For extremely high volumes of throughput, horizontal impact crushers can have double rotors fitted.

As the name implies, vertical shaft impact crushers are fed vertically but the crushing action is this time by throwing the aggregate feed in a horizontal plane. Aggregate is allowed to build up on the sides of the box, and an aggregate wall is built. This now means that aggregate is thrown against aggregate and little contact between metal wear part surfaces occurs. This is a particularly cost-effective means or providing a high-volume feed to the screens.

They are suitable for crushing higher yields of sand in flint gravels and when crushing grit-stones produce a good shape. Stone against stone crushing ratios can be higher for this type of device.

The *screening* procedure allows the production of aggregate products for component use in other products. There are many different methods of screening and, like the crushing operation can to a certain degree be tailored to suit a particular type of material.

Screening forms part of the secondary process and can involve the product, in particular the finer aggregate, being washed to allow its use if from a deposit that requires this.

As previously mentioned, some screening of the primary extracted aggregate can take place to allow for the removal of any undesired elements or to remove oversized material. This is done by the use of a set of inclined parallel bars that are vibrated; the gap width of these bars can be varied to allow for different sizes to be removed at source. This process, or type of screen, is commonly called a 'grizzly or bar' screen.

Consideration of choice of equipment must be given; this will depend on the types of aggregate sizes required and their volumes present in the deposit. For instance, if only the basic elemental aggregates of concrete production are required, then the amount and type of plant should reflect this.

Some deposits may include partially fossilized wood and lignite (low-grade coal). Specialist, dedicated plant, by means of density separation, can remove these. In the case of these contaminants (which have a detrimental effect on any finished concrete) is by the use of water.

The use of varying sized mesh apertures, which can be rounded, rectangular or square, depending on their desired function, are utilized to produce a finished aggregate product and remove any undersize or oversize from such products. This is done by attaching the screens to an inclined oscillating, vibrating device, known as the screen deck. This is a typical screen arrangement, where the angle of inclination is between 15–20° and is sometimes known as a 'Niagara' system. This action creates a backflow effect and allows the material to present to the screen, over which it is travelling, many times during the process. There are other methods of suspending screen meshes of various types, such as the arrangement mentioned above, known as 'multiple layer' decks. These greatly improve the area available and capacity of the machine. Other methods include a curved multiple-deck arrangement, as mentioned for pre-screening, to remove oversize, a set of parallel bars and, in the early days of relatively low demand and production levels on the industry a complete round barrel, fitted with curved screens. This process was excellent for producing good shaped aggregates and was known as a Trommel screen. Due to high production levels, this type of screen is rare in the modern production unit.

The efficiency of screening can depend on several factors. The rate of feed to the screen, the weather and state of wetness of the aggregate all play a part in the sizing process as well as the choice of apertures for the screen decks. Also playing a role is the choice of material from which the screen is manufactured. The original particle shape also determines the process efficiency, and this in turn can also have an influence on the type of screen deck chosen. The more rounded the particle, the more difficult the screening process may be in terms of how many particles pass the required aperture.

Screen decks can be made from woven wire (as in test sieves used within a laboratory) or from durable plastics. Wear and tear on the screen decks also depends on the abrasive nature of the aggregate being extracted. The use of woven wire decks in very wet processing may also fail due to rusting of the decks. Regular checks on the screens should be carried out in order to avoid total screen failure and the subsequent aggregate produced not meeting the required specification. They are relatively easy to maintain and replace once worn and are usually slotted or bolted into place on the main deck of the screening device. This will cover the coarse aggregate fraction of the material extracted from the deposit. The fine aggregate itself undergoes different procedures to achieve the required finish.

There are two main terms that are associated with screening efficiency: pegging and blinding. *Pegging* refers to the coarse aggregate particles becoming stuck in the screen deck matrix apertures. Excessive pegging will result in lost efficiency and will result in less area available to process the feed and in poorly graded produced aggregate. The use of tensioned wire decks here is an option to overcome this as the vibration of the screen deck over a large area can have some self-cleaning properties which can avoid this problem to a certain extent. Large decks with this arrangement, however, once failed due to wear, have a catastrophic effect on the finished aggregate. *Blinding* is usually a direct result of higher than normal levels of moisture in the aggregates being screened and the subsequent agglomerations that build up on the screen mesh. In turn, this causes a reduction in the area available for the process and an overall reduction in capacity. Inclined screens are particularly prone to this if the amplitude of vibration is insufficient and when the moisture content reaches a given level the blinding effect will take place. The use of varying size pulley wheels can be used to adjust the turn rate of the off-set cam that produces the vibration. Both the pegging and blinding effects can be avoided by regular checks on the quarry screen house and to a degree prevented by good maintenance.

The next part of the process in terms of the fine aggregate from the deposit is classification. This is done by turning the material into a fluid state by the addition of water. Injecting water at pressure causes a flow current and allows different sized particles to be extracted at certain levels within the classifier. The particles are then separated or 'classified' according to their mass, so the upward current produced by the classification process means that the fine sand will be taken from the top and, due to its greater mass, the coarse sand at the bottom.

At this stage of the process the classified sands are very wet and therefore need to undergo a process known as *de-watering*. This can either be by simply allowing the material to be stocked out and to free drain, due to a process of stock management or by means of a dedicated de-watering plant fitted after classification. In general, the finer the sand particles, the greater the amount of time that is required for the material to drain. This is due to the increased pore pressure that builds up within the finer sands and a de-watering plant breaks these pressures by using spaced strips of a material with elastomeric properties, which is fixed together with horizontal tie bars. The assembly is then vibrated and a thick layer of the sand is allowed to build up over the inclined screen deck. The subsequent vibrations of the screen cause the pore pressures and surface tension effect to be broken and the water allowed to pass through the material.

There are some variations to the method of classification; mainly the use of hydrocyclones and Archimedian screw type devices. The hydrocyclones are effectively a typical classifier arrangement and the screw types involve the movement of material up an inclined pool from the point where the sand feed enters. Again this is a type of density separation, the main disadvantage of this type of device being that generally they do not suit larger production volumes.

The amount of water removed is important for the eventual use of the product as a raw material, as excessive moisture contents can lead to difficult product quality control in end use.

6.5 Extraction and processing of limestone

The use of limestone in the production of concrete will depend on the location of the source, the availability of other sources and the type of concrete to be manufactured with the aggregate. The use of explosives to extract the primary material for processing is the fundamental difference between the extraction of sand and gravel to what are commonly called 'hard rocks' by the industry. This therefore adds an extractive additional cost to the product (due to the additional use of explosives training, supervision, storage and use). These types of aggregates are also planned for on a long-term basis and are land-won only. This factor itself attracts different levels of planning permission to extract and the process of restoration of the finished works will also require some degree of specialist treatment.

The material is extracted by explosives by using a drilling rig to make a series or pattern of holes on the top of the quarry face. Each face (and there are usually several that make up a quarry) is known as a bench and allows the material to be extracted in stages by progressively moving down a stage. An individual quarry may have several faces with each bench in a face being from 3 to 5 metres in height. It is important to divert as much of the energy from the explosives into the surrounding rock to cause a high degree of shatter and to avoid large unusable 'pop' rock that cannot be fed to the primary crushing device because of its size.

A drilling rig with a percussive action is used to drill a series of hole patterns within the quarry face on the bench to be blasted. The holes follow the profile of the face and are laid down in echelon. These are then charged with explosives, which are largely purchased from suppliers rather than manufactured on-site, called ANFO (ammonium nitrate ferrous oxide), a mix of gas oil and industrial grades of fertilizer.

If there is any part of the face which has been deemed to have softer, less resistant rocks, then these areas will not require charging with explosive and must be stemmed. This means that small aggregates are poured into the hole to avoid an escape of energy into the less resistant area and thus causing the potential risk of rock flying away from the face and resulting in damage or injury. It has been recorded that a piece of rock of some 200 mm in diameter has travelled up to 2 km from the quarry face. A method of further reducing this risk is to cover the face to be blasted with disposable netting that will catch any escaped rock.

The holes are also run down with the detonators, which are linked to the rest of the hole in the 'shot' to be 'fired'. Each detonator or 'cap', as they are sometimes known, has a specific detonation time and the plan is usually to time the operation so that the front of the blast (i.e. the part nearest the face) is fired first, followed by the rear. This takes place in milliseconds from the shot being fired.

Each blast can bring down several thousand tons of primary rock at a time and how many times a quarry blasts will depend on the volumes required to fulfil the needs. Twice to three times a week is not uncommon in an average sized quarry producing up to a million tons per year.

Any large unsuitable rock from the blast can be separated and later can be reduced by the use of a hydraulic pick fitted to a back-acting loader. (These are similar to the breaker that can be fitted to a primary crusher usually of the jaw variety: alternatively they can be sold for use as coastal or sea defence works.) Secondary, tertiary and screening processes utilize the equipment as described above (see typical layouts for these types of extraction).

6.6 Summary

The extractive minerals industry has developed rapidly from the end of the Second World War to accommodate the growth of society and the need for an improved infrastructure in a world economy. Each stage of the processes and the equipment required to extract and process raw materials into finished usable aggregates has been examined, and varying locations will require the combination of several of these processes to achieve this. The operation of an extraction site requires knowledge not only of the equipment and processes involved but also of the raw material itself, planning issues, health and safety and the environment. Other aspects not covered by this chapter are those of engineering and electrical installation and operation of the associated equipment with the extractive industry, and the author recommends a broad knowledge of these matters.

The impact of extraction of raw materials on the environment also has to be considered along with the responsible operation of sites, their use by future generations and any future recycling of other sources.

Further reading

Anon. (1999a) New Parker plant for Hereford quarry. *Quarry Management.* December.
Anon. (1999b) Mobile Plant. *Quarry Management.* August.
British Standards Institution, BS 63 Part 1, 1989.
British Standards Institution, BS 63 Part 2, 1989.
Mellor S.H. (1990) *An Introduction to Crushing and Screening.* Institution of Quarrying, Nottingham.
QPA. (1995) *Statistical Year Book* Quarry Products Association, London.
Smith, M.R. and Collis, L. *Aggregates* (2nd edn). The Geological Society, London.

7

Lightweight aggregate manufacture

P.L. Owens and J.B. Newman

7.1 Introduction, definitions and limitations

Lightweight-concrete is defined by BS EN 206-1 as having an oven-dry density of not less than 800 kg/m^3 and not more than 2000 kg/m^3 by replacing dense natural aggregates either wholly or partially with lightweight aggregates. Although excluded from this Standard this range of densities can be achieved if the concrete is made with most dense natural aggregates but in such a way that excess air is incorporated as, for example, in no-fines, aerated or foamed concrete.

Lightweight aggregate is defined in BS EN 13055 as any aggregate with a **particle** density of less than 2.0 Mg/m^3 or a **dry loose bulk** density of less than 1200 kg/m^3. These properties mainly derive from encapsulated pores within the structure of the particles and surface vesicles. For structural concrete suitable aggregates should require a low cement paste content and have low water absorption.

The most appropriate method of assessing particle density in structural concrete is to compare the density of the particle to that of typical natural aggregate, e.g. with a 'reference' particle density of 2.6 Mg/m^3. The minerals comprising the solid structure of most aggregates, whether lightweight or not, have densities close to 2.6 Mg/m^3, but it is the air within the structure of an aggregate that enables compacted structural concrete to be less than 2000 kg/m^3.

It is important to understand the difference between the two definitions of density. **Particle density** relates to the mass per unit volume of individual aggregate particles as measured by BS EN 1097-3[3]. **Dry loose bulk density** is the mass of dry particles contained within a given volume as measured by BS EN 1097-6[4]. To demonstrate this difference Figure 7.1 illustrates the relationship between these two definitions of density in relation to particle shape, texture and interstitial voids. It is thus apparent that those aggregates with the lowest interstitial void space are those which give the lowest concrete densities.

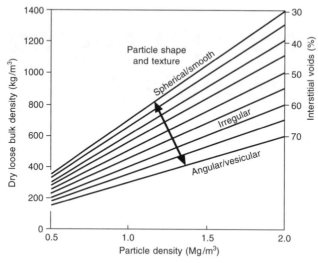

Figure 7.1 Relationships between dry loose bulk density and particle density for manufactured aggregate.

7.2 Lightweight aggregates suitable for use in structural concrete

For structural concrete, the requirements for lightweight aggregate are that it:

- has a strength sufficient for a reasonable crushing resistance (BS EN 13055)
- enables concrete strengths in excess of 20 MPa to be developed, and
- produces compacted concrete with an oven-dry density in the range 1500–2000 kg/m^3.

For concrete using fine aggregate from natural sources it is the particle density of the coarse aggregate that has the greatest influence on the oven-dry density of compacted concrete. To achieve lightweight concrete the particle density of the coarse aggregate should be less than about 0.65 Mg/m^3 at the lower end of the density range or greater than about 1.85 Mg/m^3 at the higher end. Typically, an aggregate particle density of 1.3 Mg/m^3 would produce concrete with a fresh density of about 1750 kg/m^3.

Traditionally, manufacturers of lightweight aggregates have been constrained by not only the limitations of the raw material and the method of processing but also the requirements of the market. The properties of lightweight concrete, and hence lightweight aggregate, can be exploited in a number of ways from its use as a primarily structural material to its incorporation into structures for the enhancement of thermal insulation. This variety of purpose is recognized by RILEM/CEB who proposed the classification shown in Table 7.1.

This chapter relates to the manufacture of aggregates for concretes within Class I (i.e. structural lightweight concrete).

Table 7.1 RILEM/CEB classification of concretes (RILEM, 1978)

Class Type	I Structural	II Structural/Insulating	III Insulating
Compressive strength (N/mm^2)	>15.0	>3.5	>0.5
Coeff. of thermal conductivity (W/m.°K)	n/a	<0.75	<0.30
Approx. density range (kg/m^3)	1600–2000	<1600	<<1450

For aggregate manufacture, the availability of the mineral resource will be crucial since the lower the finished aggregate **particle density**, the less will be the rate at which the resource will be depleted for the same volume of production. This creates an obvious conflict of interest – does a producer develop a lightweight aggregate for a market which depletes the resource at a greater rate or for a speculative market with the possibility of a smaller return on capital? This dilemma has caused more lightweight aggregate producers to fail than in any other segment of the aggregate industry.

7.3 Brief history of lightweight aggregate production

The history of lightweight aggregate production from natural sources dates back to pre-Roman times and continues today with porous rocks of volcanic origin. However, the sources are limited to regions of volcanic activity.

From around the end of the nineteenth century, with the development of reinforced concrete and the rarity of natural porous aggregate deposits and their non-existence in most developed countries, research for the manufacture of aggregates commenced. In Europe, in the early part of the twentieth century development concentrated on foaming blastfurnace slags since steel production was basic to the industrial infrastructure. However, it was not until the early 1970s that significant developments in pelletizing and expanding blastfurnace slags took place, so that today a slag-based aggregate with a smoother non-vesicular surface, more adaptable for structural concrete, is produced.

It was not until about 1913 that research in the USA revealed that certain clays and shales expanded when fired. In about 1917 this led to the production in Kansas City, Missouri in a rotary kiln of a patented expanded aggregate known as 'Haydite' which was used in the construction of the USS *Selma*, an ocean-going ship launched in 1919. There followed in the USA the development of a series of aggregates known as 'Gravelite', 'Terlite', 'Rocklite', etc. In Europe, however, it was not until 1931 that the manufacture of LECA (lightweight expanded clay aggregate) commenced in Denmark. Thereafter, developments rapidly spread to Germany, Holland and the UK Rudnai (1963). The principle of expanding suitable argillaceous materials, such as the geologically older forms of clay (e.g. shale and slate), has been exploited in the UK since the 1950s with the products being marketed as 'Aglite', 'Brag', 'Russlite' and 'Solite'. All these companies failed for one reason or another, mainly due to the inaccessibility of the market, non-homogeneity of the feedstock, undesirable gaseous and particulate emissions and the high cost of production, **but never on the technical performance of the aggregate in concrete**.

In the 1950s the UK Building Research Establishment (BRE) developed the technology for the production of a high-quality lightweight aggregate based on pelletized pulverized-fuel ash (PFA) resulting from the burning, generally in power stations, of pulverized bituminous or hard coals (Cripwell, 1992). PFA is the residue of contaminants in hard coals that are present as the result of erosion of natural minerals which sedimented into the coal measures as they formed. Thus, there is a connection between PFA and other argillaceous minerals, except that most of the PFA has already been subjected to temperatures in excess of 1250°C, which vitrifies and bloats some of the particles which are known as cenospheres. Two construction companies became involved in the exploitation of PFA and the knowledge obtained by BRE. Cementation Ltd operated at Battersea Power Station with two shaft kilns which failed and John Laing & Co. Ltd at Northfleet with a sinter strand which became the most successful method in the UK of producing structural grades of lightweight aggregate from PFA. It was, and is, marketed under the trade name of 'Lytag' and has an ability to reduce the fresh density of concrete to about 1750 kg/m^3 when using Lytag fines and about 1950 kg/m^3 with natural fines. These fresh densities are equivalent to oven-dry densities of about 1575 kg/m^3 and 1825 kg/m^3 respectively.

For the production of structural concrete with an oven-dry density of less than 1800 kg/m^3, a German manufacturer has produced from about the early 1970s a spherical expanded shale aggregate ('Liapor') in a range of particle densities from 0.80 to 1.70 Mg/m^3. This represents one of the most significant advances in lightweight aggregate manufacture since the aggregate particle can be designed to suit a range of oven-dry concrete density requirements, thereby giving greater versatility of application.

Other developments are taking place in lightweight aggregate manufacture, such as the production of hybrids using PFA or other pulverized materials and suitable argillaceous minerals (clay, shales and slate). There are also aggregates termed 'cold bonded' which are mixtures of PFA with lime or Portland cement. Although the manufacture of these aggregates is now becoming more successful, their use is more appropriate to the production of masonry than to structural concrete.

7.4 Manufacturing considerations for structural grades of lightweight aggregate

7.4.1 Investment

For any lightweight aggregate, considerable investment in manufacturing plant is required, sustainable quantities of appropriate resource material must be available and there must be a market. In the USA, most cities are based on the principle of high-rise development which in many cases leads to the use of lightweight aggregate concrete. Europe, with its more historic traditions and older infrastructure, has lagged behind. However, more consideration is being given to the conservation of land-based mineral resources and it is more difficult to obtain permission for mineral extraction. The introduction of taxes on natural materials, which will increase, and the ever-increasing cost of transport, etc. will provide a further impetus for investment in aggregate manufacture.

7.4.2 Resource materials

The most important asset for any lightweight aggregate manufacturer is a sustainable source of raw material in a form and state ready for immediate use. Manufacturers using either PFA or molten slag have immediate advantages compared with those using other resource minerals. However, the limitations with these materials are that the process of sintering fuses the PFA particles in such a way that it densifies the agglomerate, while the difficulty of entraining air into molten slag limits the **particle** density to about 1.75 Mg/m^3.

Alternatively, an aggregate based on argillaceous materials such as clay, shale or slate can have its density varied by the manufacturing technique. The lighter the aggregate, the lower the demand for resources whereas for the stronger and denser aggregates the resource is depleted at a greater rate. A solution is to make a hybrid aggregate based on PFA mixed with between 25 to 75 per cent by mass of clay (British Patent Specification No. 2218412) which can be heated or fired to make an aggregate with a particle density of, say, 1.35 ± 0.5 Mg/m^3

7.4.3 Processes of lightweight aggregate manufacture

Most manufacturing processes for lightweight aggregates, with the exception of processes using blastfurnace slag, have been limited to the use of either a sinter strand or a rotary kiln. In instances where the aggregate particle is formed as a friable fresh pellet before firing the sinter strand is preferred. Where the form of the fresh pellet is cohesive and its shape can be retained, the rotary kiln produces the most spherical particle with the most impermeable surface.

7.4.4 Production techniques

The various production techniques rely either on agglomeration or expansion (bloating). Agglomeration takes place when some of the materials melt at temperatures in excess of about 1100°C and the particles that make up the finished aggregate are bonded together by fusion. Alternatively, expansion develops when either steam is generated, as in the case of molten slag, or a suitable mineral (clay, shale or slate) is heated to fusion temperature, at which point pyroplasticity occurs simultaneously with the formation of **gas** which bloats the aggregate.

When appropriate argillaceous materials are heated by firing to achieve appropriate expansion, the resource mineral should contain sufficient gas-producing constituents and reach pyroplasticity at the point of incipient gas formation. Gas can be developed by a number of different reactions, either singularly or in combination, from the following:

(a) volatilization of sulphides from about 400°C
(b) decomposition of the water of crystallization from clay minerals at approximately 600°C
(c) combustion of carbon-based compounds from approximately 700°C
(d) decarbonation of carbonates from approximately 850°C
(e) reaction of Fe_2O_3, causing the liberation of oxygen from about 1100°C.

Most argillaceous materials that are suitable become pyroplastic at between 1100°C and 1300°C. However, depending on the actual source of the material and its chemical composition, the temperature at which bloating for each material becomes effective is within a relatively small range, usually about ±15°C. At this point the bloated material has to be removed immediately from the firing zone and cooled quickly to 'freeze' the particle at the desired degree of bloat, otherwise it will continue to expand. Ultimately, if expansion is uncontrolled, the pore wall becomes too thin and there is insufficient resistance to crushing and the particle will not be strong enough to resist the stresses achieved within structural concrete.

A principle of success with all lightweight aggregate manufacture is homogeneity of the raw material source, as variability inevitably causes fluctuations in manufacture and the finished product. To emphasize this present-day aggregate manufacturers have gone to considerable lengths to ensure homogeneity of the raw material.

7.5 Production methods used for various lightweight aggregates

7.5.1 Foamed slag aggregate

In this process molten blastfurnace slag at more than 1350°C is usually poured onto a foaming bed consisting of a large number of water jets set in a concrete base. The water immediately converts to steam on contact with the molten slag and penetrates into the body of the material, at which point the steam becomes superheated. Owing to the rapid expansion that then takes place, the slag foams to form a cellular structure and vesicular surface texture. Alternative methods of expansion include spraying water onto the molten material when it is being tapped from the blastfurnace so that the material is cooled rapidly, with steam becoming entrapped within the structure of the particle. At the completion of foaming the slag is removed and stockpiled, from where it is subsequently crushed and graded to size. The aggregate produced is very angular with an open vesicular surface texture (Figure 7.2).

Figure 7.2 Typical foamed slag particles.

7.5.2 Pelletized expanded blastfurnace slag aggregate

The process of slag pelletization was developed in the early 1970s in order to overcome environmental concerns associated with the production of foamed blastfurnace slag on open foaming beds or pits. Not only does the slag pelletization process overcome these concerns it also produces an aggregate with a closed surface.

To manufacture this aggregate, molten blastfurnace slag at a temperature of about 1400°C passes through a refractory orifice or 'block' to control the rate of flow. It then flows onto an inclined vibrating plate with water running down its surface. The vibration of the plate breaks up the slag flow and traps water which immediately vaporizes and expands the slag.

More water is sprayed onto the surface of the slag at the end of the vibrating plate. This enables gas bubbles to form in the body of the slag, creating further expansion while also chilling its surface. After being discharged from the vibrating plate the expanded globules of semi-molten slag are fed onto a horizontal rotating drum fitted with fins which project the material through a water mist. The trajectory of material is such that the slag forms irregular and more spherical pellets which are sufficiently chilled to avoid agglomeration when they come to rest. After pelletization the material is removed, allowed to drain and, finally, screened to the required size.

The process produces a finished product that comprises semi-rounded pellets with a smooth surface encasing a glassy matrix and a discrete cellular structure, which is essentially non-absorbent (Figure 7.3)

Figure 7.3 Typical 'Pellite' particles.

7.5.3 Sintered pulverized fuel ash (PFA) aggregate

This aggregate is produced from PFA which is a powder by-product of pulverized bituminous coal used to fire the furnaces of power stations. Suitable PFA of not more than about 8–10 per cent loss on ignition which results from unburnt carbon in the form of coke (char), is first homogenized in bulk in its powder form. Once homogenized, it is then conditioned through a continuous mixer with about 12–15 per cent of water and, as necessary, an amount of fine coal is added to make up the fuel content to about 10 per cent of the dry mass of the pellet to enable it to be fired. This conditioned mixture of PFA is then fed at a controlled rate onto inclined and revolving pelletizing dishes. The size and degree of compaction of the formed 'green' pellets depend on the inclination and speed of rotation of the dish, the rate of addition of the conditioned PFA as well as a further amount of water spray. The formed pellets discharge from the pelletizer at a diameter of about 12–14 mm. Without any further treatment these pellets are conveyed to the sinter strand where they are fed by spreading to form an open-textured and permeable bed to the width and depth of the sinter strand. The sinter strand is a continuously moving conveyor comprising a series of segmented and jointed grates through which combustion air can be drawn to fire the pellets and combustion gases exhausted. Once on the bed, the sinter strand immediately carries the pellets under the ignition hood that fires the intermixed 'fuel'. The chemical composition of PFA resembles that of clay but, unlike clay, as the PFA has already been fired, no pre-drying or pre-heating of the pellets is necessary, as the pellet is able to expire the water as vapour and combustion gases without incurring damage to the particles. Once ignited at about 1100°C, and as the bed moves forward, air for combustion is drawn by suction fans beneath the grate. The process is controlled to prevent the particles of PFA becoming fully molten so that (a) they coagulate sufficiently to form an aggregate and (b) the aggregate particles are only lightly bonded to each other. The correct amount of coagulation within the pellets is obtained by varying both the speed of the sinter strand and the amount of air drawn through the bed of pellets.

The finished product on the sinter strand is a block of hard brick-like spherical nodules, lightly bonded by fusion at their points of contact. As the sinter strand reaches the end of its travel and commences its return to the feeding station a segment of the bed of the finished product is discharged into a breaker. This action separates the aggregate pellets prior to grading.

While the surface and internal structure of the finished pellet (Figure 7.4) is essentially 'closed' it contains encapsulated interstices between the coagulated PFA particles. While these interstices are minute they are penetrable by about 20 per cent moisture but eventually 'breathe' sufficiently to allow any water to evaporate even when encased in concrete.

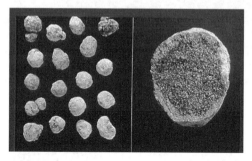

Figure 7.4 Typical sintered PFA particles.

7.5.4 Expanded clay aggregate

Expanded clay aggregates are manufactured in rotary kilns each consisting of a long, large-diameter steel cylinder inclined at an angle of about 5° to the horizontal. The kiln is lined internally in the firing zone with refractory bricks which, as the kiln rotates, become heated to the required temperature and 'roast' the clay pellets to achieve the required degree of expansion. The length and configuration of the kiln depends in part on the composition of the clay and length of time it takes to 'condition' the clay pellets in the pre-heater to reach a temperature of about 650°C to avoid them shattering before becoming pyroplastic.

The clay is dug and, to eliminate natural variability, is usually partly homogenized by layering into a covered stockpile with a spreader before it is removed from the stockpile by scalping with a bucket conveyor. In the process, a high degree of blending is achieved. The clay is prepared by mixing thoroughly to a suitable consistency before pelletization. This latter process aims to produce smaller pellets than are required for the finished product as their volume can be increased by two to six times. The prepared pellets are fed into the kiln which can be considered to be in three segments, namely, drying and preheating at the higher end and firing and then cooling at the lower end. During the progress of the prepared material through the kiln, the temperature of the clay pellets gradually rises until expansion occurs. The expanded product is discharged from the firing zone as soon as possible for cooling to freeze the particles at the desired degree of expansion. Cooling takes place in either a rotary cooler or a fluidized bed heat-exchanger. The finished product is graded and, if necessary, crushed to particle sizes less than 16 mm. While the particle density can be varied depending on the range of temperatures at which expansion takes place, the mean expansion temperature is about $1175 \pm 50°C$. This varies for different clays but most manufacturers are limited to expandable clays which have a confined range of bloating temperatures. In some cases, the range is less than 25°C between non- and full expansion. In instances such as this, the scope is limited for the manufacture of intermediate density grades. However, manufacturers of such aggregates, by requiring to optimize their production, usually have preferences for aggregates with lower **particle** densities, in the range 0.4–0.8 Mg/m^3, as this tends to conserve the resource which becomes depleted at a greater rate at higher particle densities. Thus, there is a greater attraction to produce aggregates for thermal insulation than for structural applications. However, for the lower density structural concrete, i.e. 1300–1600 kg/m^3, these aggregates are the most suitable. As shown in Figure 7.5, the surface texture is closed and smooth with a 'honeycombed' or foamed internal structure, where the pores are not interconnected.

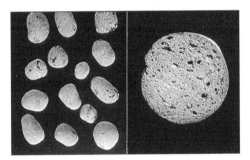

Figure 7.5 Typical expanded clay particles.

7.5.5 Expanded shale and slate aggregate

Shale

Shale, a low-moisture content soft rock, is quarried, transferred to blending stockpiles before it is reduced by primary crushers and dry-milled to a powder of less than 250 μm. This powder is homogenized and stored ready for pelletization in manner similar to that used for making aggregate from PFA except that no fuel is added. However, after the pellets have been produced to the appropriate size, which depends on the expansion required, they are compacted and coated with finely powdered limestone. The resulting pellets are spherical with a 'green' strength sufficient for conveying to a three-stage rotary kiln consisting of a pre-heater, expander and cooler. Unlike other aggregates produced from argillaceous materials, the feedstock is reduced to a powder and then reconstituted to form a pellet of predetermined size. The expansion (bloating) is controlled during kilning to produce an aggregate of the required particle density. Different particle densities are produced by controlling the firing temperature and the rotational speed of the kiln. The coating of limestone applied to the 'green' pellet increases the degree of surface vitrification which results in a particle of low permeability. This product gives versatility to the designer for pre-selecting an appropriate concrete density. As Figure 7.6 shows, while the particle shape and surface texture of the aggregate remain essentially the same, the internal porosity can be varied according to the bloating required for the specified density.

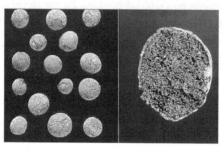

Figure 7.6 Typical expanded shale aggregate.

Slate

Slate, a hard argillaceous rock, is first reduced to about 12–15 mm and is then fed into a three-stage kiln consisting of a pre-heater, expander and cooler which can be a pond of water. The firing temperature is about 1150°C at which the laminated platelets of slate become pyroplastic and the gases released cause the particles to expand to form an almost cubical shape. The process, which must, again, be very carefully controlled, reduces the particle density from that of the slate, about 2.70 Mg/m³, to a pre-determined density between 0.65 and 1.25 Mg/m³. The finished product is crushed and graded to a maximum size of 25 mm and the surface texture of the finished particle is coarse and rough. However, the surface is sufficiently closed, owing to its vitrified nature, such that internally it has an extremely low water absorption.

7.6 The future

Research on structural concrete has established that concrete made with appropriate lightweight aggregates does not have many of the deficiencies in performance compared to concrete made with natural aggregates. As the need increases for alternatives to aggregates from natural sources, there are opportunities for capitalizing on the resources that exist, for which there is currently no accepted use in their present form. For instance, it is estimated that there are some 200 million tonnes of 'stockpiled' PFA in landfill in the UK alone. In other EC countries landfilling of PFA is now penalized for environmental protection by taxation of the power companies. Although PFA has been used in the UK for the manufacture of aggregate, the finished product has a single nominal density which has limited its wider application in structural concrete. This, together with the limitations on the amount of excessive and variable quantities of residual fuel that the PFA sometimes contains, restricts full exploitation of the technology of conversion to a structural aggregate.

Expanded clays have their own limitations, mainly because of restrictions placed on the availability of suitable material. In 1992 a plant for the production of expanded clay aggregate in the UK was closed due to the refusal of planning permission for the winning of otherwise accessible resources.

Other alternatives exist for combining the technologies used in the manufacture of aggregates from both clay and PFA to produce products identical in performance to those of slate or shale. The amount of clay used can be as low as 10 per cent but can be as high as 80 per cent by mass when particle densities lower than 0.65 Mg/m^3 are required. The loss on ignition in the PFA is less restrictive as up to 20 per cent can be accommodated but it must be controlled to low variability. The advantages to manufacture are that the aggregate can be fired in a rotary kiln without a pre-heater and the cooler can be a fluidized bed heat exchanger, both of which reduce the size of the plant to the essential rotary kiln. Particle densities of the finished product can be more easily varied as the range of temperatures is controlled to produce the density required. The particle is spherical with a low water permeability and a smooth surface texture.

The sinter strand, can use any argillaceous material, for instance clay or the considerable stockpiles of 'soft' slate that was produced as waste when quarried (e.g. in North Wales, while the spoil and tailings produced from mining coal are, in theory, potential raw materials for lightweight aggregate production). However, the volatiles in mine tailings, etc. present a concern for air pollution control while clay, shales and slate containing low amounts of volatile products are most suitable. However, for lightweight aggregates to be successful in the future, pulverization and reconstitution of the shale or slate into pellets must be an integral part of the manufacturing process. Producing aggregates on a sinter strand results in a less spherical particle but with the advantage of a closed surface.

For increasing the suitability of mine 'tailings' as an aggregate feedstock, the removal of volatiles such as naphtha, sulfur and bituminous by-products would be essential before pulverization and homogenization of the resultant 'clean' feedstock prior to pelletization.

The most significant development is the use of resources from so-called 'waste' materials for the manufacture of lightweight aggregates (Cheeseman et al., 2000). Such resource materials include residues from the incineration of municipal waste (incinerator bottom ash (IBA) and air pollution control residues), sewage sludge, waste clay, etc. This will address two major problems, namely, the increasing difficulty of waste disposal due to EU restrictions on landfill and the provision of aggregate which is becoming more difficult to obtain for a variety of reasons, particularly in conurbations where there is reduced access to local resources. The use of state-of-the-art efficient thermal processing techniques and the financial benefits resulting from the use of 'wastes' will allow the price of the manufactured aggregate to be reduced to a level at which lightweight concrete of a given grade will be no more, and possibly less, expensive than normal dense concrete but with enhanced properties (Owens and Newman, 1999).

7.7 Conclusions

The technology is available for successful lightweight aggregate manufacture. However, to produce a generic and consistent aggregate the feed stocks used must be statistically homogenized before pelletization. The manufacture of aggregates in the UK from slate waste in North Wales and from minestone in East Kent, Derbyshire and Scotland, all failed, not because of technical deficiencies of the aggregate itself, but because of difficulties with the preparation of an inconsistent raw material, its process control and emissions from the plants. Alternatively, aggregates have been produced from slag, a highly controlled by-product of iron production, as well as from PFA. Both have succeeded since considerable care is taken to remove variability from the raw materials to enable production to continue with as little 'manual' interference as possible.

Another concern with all lightweight aggregate manufacture is to match demand with supply. Significant investment will not be made until it becomes common practice to use lightweight concrete for structural purposes both below, as well as above, ground level. For example, One Shell Plaza in Houston (Khan, 1969) used lightweight concrete in the foundations and basement, proving that engineers can not only utilize the advantages of lightweight concrete, but can also make it commercially advantageous for owners of all types of structures.

References

BS EN 206-1: Concrete: Part 1: Specification, performance, production and conformity.

BS EN 13055: Part 1: Lightweight aggregates: Part 1: Lightweight aggregates for concrete, mortar and grout.

BS EN 1097-3, Tests for mechanical and physical properties of aggregates: Part 3: Determination of loose bulk density and voids.

BS EN 1097-6, Tests for mechanical and physical properties of aggregates: Part 6: Determination of particle density and water absorption (Annex C).

BS EN 13055: Part 1: Lightweight aggregates: Part 1: Lightweight aggregates for concrete, mortar and grout (Annex A: Determination of crushing resistance).

Cheeseman, C., Newman, J. and Owens, P. (2000) Why waste waste? A blueprint for generating products from waste in London. Institute of Wastes Management, Millennium Competition, Award Winning Entries, IWM, October.

Concrete Society (1978) Structural Lightweight Aggregate Concrete for Marine and Off-shore Applications, Report of a Concrete Society Working Party, Technical Report No. 16, May.

Cripwell, J.B. (1992) What is PFA? *Concrete*, **26**, No. 3; May/June, 11–13.

Khan, F.R. (1969) Lightweight concrete for total design of One Shell Plaza. ACI Special Publication SP29, Lightweight Concrete Paper SP29-1.

Owens, P.L. and Newman, J.B. (1999) Increasing the environmental acceptability of new energy from waste plants by integration with cost effective concrete aggregate manufacture. *IWA Scientific & Technical Review*, Institute of Wastes Management, Nov. 21–26.

RILEM (1976) *Functional classification of lightweight concretes*, Recommendation LC2, 2nd edn.

Rudnai, G. (1963) *Lightweight Concretes*. Akademiar Kiado Publishing House of the Hungarian Academy of Sciences, Budapest.

8

The effects of natural aggregates on the properties of concrete

John Lay

8.1 Aims and objectives

The purpose of this chapter is to explain how the properties of natural aggregates influence the properties of fresh and hardened concrete. It describes British and European Standard tests for aggregates and emphasizes the importance of correct sampling and describes the interpretation of aggregate test results. The chapter also considers the likely effects of impurities associated with natural aggregates.

8.2 Brief history

One of the earliest known concretes was placed in 5600 BC on the banks of the river Danube (Stanley, 1979). It contained sand and gravel aggregate, bound with lime. Ancient concretes have also been found consisting of broken rock, sand and lime. For many centuries, transport constrained the choice of aggregates. In almost all cases, local materials were used. Even so, differences between the performance of different rocks in concrete have been known since Roman times (Sims and Brown, 1996). In the modern age, it has become more practical to transport aggregates and specifications have grown up to assess materials. Early specifications contained subjective clauses, such as the requirement that aggregates should be clean, hard and durable. The phrase survives to this day in the Specification for Highway Works (Highways Agency, 1998). In the latest version of BS 882, the British Standard Specification for Aggregates from Natural Sources for Concrete, the requirements are expressed quantitatively and concisely (BSI, 1992). BS EN 12620, the European Standard for Aggregates for Concrete, also defines requirements quantitatively (CEN, 2002).

8.3 Introduction

Aggregate is the main constituent of concrete. Aggregate properties do influence concrete properties, but by and large they do not control the performance of the concrete. The essential requirement is that the aggregate remains stable within the concrete in its exposure conditions. The choice of aggregates used is often constrained by haulage costs.

There are three main reasons for mixing aggregate with cement paste to form concrete, rather than using cement paste alone. The first and oldest reason is that aggregate is cheaper than cement, so its use extends the mix and reduces costs. Second, aggregate reduces shrinkage and creep, giving better volume stability. Third, aggregate gives greater durability to concrete. Many deterioration processes principally affect the cement paste.

There are clear economic and technical reasons for using as much aggregate and as little cement as possible in a concrete mix. Since the economic arguments are obvious, the rest of this chapter will consider technical aspects of the relationship between the properties of aggregate and the properties of concrete.

8.4 Classification

Petrological classification alone is an inadequate guide to aggregate performance. While petrographic descriptions of aggregates are interesting and useful in assessing quality, they do not provide any more than general guidance on performance in concrete. The DoE mix design method, *Design of normal concrete mixes* (Teychenné *et al.*, 1988), takes into account only two types of aggregate: crushed and uncrushed. The nominal maximum size of the coarse aggregate is considered. Fine aggregate is characterized by the proportion passing a 600 μm test sieve. The particle density of the combined aggregate is also required.

Mix design methods based on trial mixing are likely also to take account of geological type, bulk density (see section 1.10) and 10 per cent fines value (see section 1.9).

8.5 Sampling

Although it is not an aggregate property, correct sampling, sample reduction and sample preparation are essential to the determination of all aggregate properties. The results of tests are meaningless if sampling procedures are not strictly observed. The aim of sampling is to obtain a test portion test that is representative of the average quality of the batch

(BSI, 1989a). A batch is a definite quantity of aggregate produced under conditions which are presumed to be uniform (BSI, 1989b).

British and European Standards give detailed procedures for obtaining and reducing samples (BSI, 1989a; CEN, 1996, 1999). The process for obtaining samples involves taking increments at random from the batch of aggregate and combining the increments to form a bulk sample. Table 8.1 shows the minimum number of increments required to form a bulk sample according to the requirements of BS 812: Part 102. The European Standard, BS EN 932-1, requires that the quantity of sampling increments taken to form the bulk sample be based on previous experience. Research funded by the European Union indicates that a minimum of fifteen increments is necessary in most cases (Ballmann et al, 1996). BS EN 932-1 recommends that the minimum mass of bulk sample should be six times the square root of the maximum particle size (in mm) multiplied by the loose bulk density (in Mg/m^3). This leads to similar bulk sample masses to those shown in Table 8.1.

Table 8.1 Minimum number of sampling increments (BSI, 1989a)

Nominal size of aggregate (mm)	Minimum number of sampling increments		Approximate minimum mass for normal density aggregate (kg)
	Large scoop (at least 2 litres)	Small scoop (at least 1 litre)	
28 and larger	20	–	50
5 to 28	10	–	25
5 and smaller	10 half scoops	10	10

Sample reduction is usually necessary to provide a laboratory sample from the bulk sample, and then test portions from the laboratory sample. The principle is again to ensure that the reduced sample is representative of the larger sample, and therefore representative of the batch. Quartering and riffling are suitable methods for sample reduction.

Quartering (which in fact halves the sample) involves mixing the sample and dividing it into approximately equal quarters. A pair of diagonally opposite quarters is retained and combined to form the reduced sample. The process can then be repeated to further reduce the sample mass if necessary.

Riffling involves passing the sample through a sample divider consisting of a row of channels. Alternate channels lead to two separate boxes. When the sample is passed through the channels, it is separated into two halves, one half being retained in each box (Figure 8.1).

For some tests, a further sample preparation stage is required. For example, a number of tests for mechanical properties are carried out on the fraction of the aggregate passing a 14 mm test sieve but retained on a 10 mm test sieve. For some chemical tests, the sample is ground to a fine powder and further reduced before analysis. Details of these final sample preparations are included in the test procedures.

8.6 Grading

Grading is the common term used to mean the particle size distribution of the aggregate. Surprisingly, variations in grading have only minor effects on hardened concrete properties.

(a) Quartering using a palette knife   (b) Quartering using a quartering cross

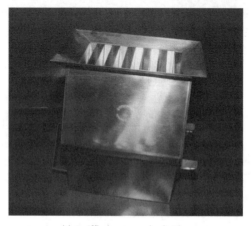

(c) A riffle box sample divider

Figure 8.1 Methods of sample reduction.

Figure 8.2 (Miller, 1978) shows the effect of changes in a coarse aggregate grading on compressive strength at constant workability. All concrete was batched to 50 mm slump. It is interesting to note that the BS 882 limits for the percentage passing the 10 mm sieve

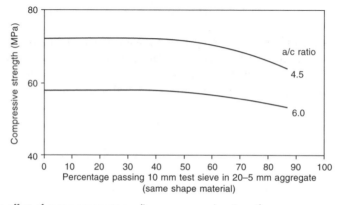

Figure 8.2 The effect of coarse aggregate grading on compressive strength.

in 20–5 mm graded aggregate are 30–60 per cent. The graphs are virtually flat lines in this range. Figure 8.3 (Ryle, 1988) shows the effect of changes in sand grading at constant workability. Within the typical range of sand gradings, the effect is fairly small. Again, it is interesting to note that the peaks of the curves fall more or less at the mid-point of the overall BS 882 limits for the percentage of sand passing the 600 µm test sieve. The influence of grading on fresh concrete is much greater. Workability, cohesion, handling, compaction, bleeding and finishing can all be affected. Grading limits have usually been derived to maintain consistency with available materials.

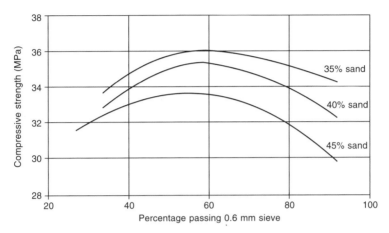

Figure 8.3 The effect of sand grading on compressive strength.

The theoretical objective of aggregate grading is to provide the maximum relative volume occupied by the aggregate, while maintaining the lowest aggregate surface area. This is because the space in the concrete mix not occupied by aggregate must be filled with cement paste, and because the surface of all the solids in the mix must be wetted. In practice, a compromise between these demands, which in some ways conflict, must be achieved. Modern mix design methods mean that aggregates with quite a range of gradings may be used (Dewar, 1983). However, uniformity and consistency of materials are still needed.

Particle size distribution is determined by drying test portions of aggregate and passing them through a nest of sieves of decreasing aperture (BSI, 1985). The amount of material retained on each sieve is weighed and the masses used to calculate the percentage of the sample that has passed each test sieve. It is preferable to wash the test portions free of fines and redry them before sieving. This breaks down any agglomerated particles in the sample (BSI, 1985). It is important not to overload the test sieves because they may blind and retain particles finer than the sieve aperture.

Grading limits in British Standards are specified in the form of tables. For each aggregate product, test sieves are listed alongside allowable ranges for the percentage of material passing that sieve. In European Standards, the aggregate producer is required to declare a grading and remain within a specified tolerance of the declared grading. There are also some broad overall limits for products. The European Standard approach is intended to encourage consistency.

8.7 Maximum size of aggregate

From the point of view of economy and minimum surface area, it would appear that the larger the maximum size of the aggregate, the better. However, larger aggregate sizes introduce problems in handling and placement, especially in reinforced concrete. Large particles may be unable to pass between congested reinforcement bars or may exceed depth of cover to reinforcement. Another problem with larger materials is that, while the reduced surface area needs less wetting, it also provides less area for the paste to bond to the aggregate. This can lead to reduced cohesion. It also results in a higher bond stress, and may lead to weaker concrete. Larger particles of crushed rock are far more likely to contain defects and will often be weaker than smaller particles of the same material (Neville, 1995). Large sizes are rare in gravel deposits. In general, lean mixes benefit from a large aggregate size (up to 150 mm if available). For normal strength concretes, optimum maximum size usually lies between 20 mm and 40 mm. For very high strength concretes, 10 mm aggregate is best, probably due to improved bond (Neville, 1995). (See volume 3, Chapter 3 in this series for more detail.)

8.8 Aggregate shape and surface texture

The shape of aggregate particles can affect water demand and workability, mobility, bleeding, finishability and strength. Equidimensional shapes are preferred. Flaky or elongated particles tend to be detrimental. Rounded particles tend to give better workability than crushed or angular particles. Producers can exercise some control over particle shape through the processing of the aggregate.

The surface texture of aggregate particles can range from glassy, through smooth, granular, rough and crystalline, to honeycombed. The texture is only really an issue where flexural strength is important, or for very high-strength concretes. In both cases, rougher textures give greater strengths, all other things being equal, because the aggregate–cement paste bond is improved.

Figure 8.4 (Ryle, 1988) shows the relationship between cement content and compressive strength for a crushed rock, a land-won flint and a marine flint. At higher cement contents, the angular and rough crushed rock produces a higher-strength concrete because of improved bond. At low- and medium-cement contents, the smooth and rounded marine gravel exerts a lower water demand for the same workability than crushed rock because there is less surface to wet.

The most common shape test is flakiness index. Coarse aggregate particles are presented to a special gauge (Figure 8.5). If a particle's least dimension is less than 60 per cent of its mean dimension, it passes through the gauge and is classed as flaky (BSI, 1989b). Flakiness index is defined as the percentage by mass of flaky particles in the sample. The European Standard flakiness test operates on a similar principle, using sieves with rectangular (rather than square) apertures. Confusingly, this test classes a particle as flaky if its least dimension is less than 50 per cent of its upper sieve size. A test survey has established an approximate relationship between the two indices (Eurochip, 1996). The survey shows that the British Standard requirements of maximum flakiness index 50 for uncrushed gravel and 40 for crushed rock and crushed gravel are equivalent to European Standard categories of 50 and 35 respectively.

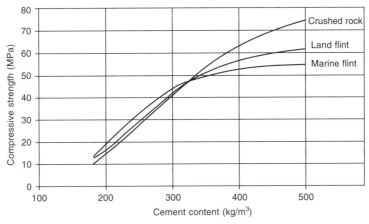

Figure 8.4 A relationship between cement content and compressive strength.

Flaky particles (left)

Elongated particles (right)

(a)

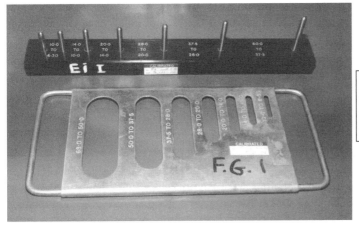

Elongation gauge (above)

Flakiness gauge (below)

(b)

Figure 8.5 Flakiness and elongation.

Other shape tests include elongation index and shape index. A particle is classed as elongated (using a gauge) if its longest dimension is greater than 180 per cent of its mean dimension. No European Standard for elongation index is planned. Shape index is defined as the proportion of particles in a sample whose greatest dimension is more than three times their least dimension. A slide gauge is used to assess particles. A European Standard test for shape index exists, but flakiness index will be the reference method for the determination of shape.

Where an assessment of surface texture is required, a qualitative description is usually given. Quantitative methods tend to be laborious and not widely used.

8.9 Aggregate strength

Aggregate strength is rarely a problem. The strength of concrete cannot exceed the strength of the aggregate. In practice, concrete grade is likely to be much less than the strength of the aggregate, because stress concentrations are generated when the concrete stresses are shared by the aggregate and cement paste. Weak aggregates may break down during mixing, handling and compaction.

Within British Standards, the mechanical properties of aggregate can be assessed from 10 per cent fines value or aggregate impact value. In the 10 per cent fines test, a load is applied to the 14–10 mm fraction of the aggregate. The load required to partially crush the sample so that 10 per cent of it passes a 2.36 mm test sieve is called the 10 per cent fines value. In the aggregate impact value test, the 14–10 mm fraction of the aggregate receives fifteen blows from a drop hammer. The aggregate impact value is the proportion of the sample that passes a 2.36 mm sieve after the hammer impacts.

European Standards will not contain the 10 per cent fines test or the aggregate impact value test (although a different impact test, the German Schlagversuch test, has been included). The reference mechanical test will be the Los Angeles test. The 14–10 mm fraction of the aggregate is placed in a steel drum with eleven large steel balls. The drum is rotated 500 times, partially fragmenting the sample. The Los Angeles value is the proportion of the sample passing a 1.6 mm sieve after fragmentation.

There is no direct correlation between 10 per cent fines value or aggregate impact value and Los Angeles value. BS 882, The British Standard Specification for Aggregates from Natural Sources for Concrete, specified 10 per cent fines values or aggregate impact values based upon the end use of the concrete. Because the Los Angeles value does not correlate with the required properties, the UK National Guidance (BSI, 2003) to 12620, Aggregates for Concrete, recommends a single value for normal concrete (Table 8.2).

8.10 Aggregate density

Loose bulk density reflects the grading and shape of aggregate, since it measures the proportions of voids in a given volume of aggregate. Its main use is as a consistency check on materials. It is also used to convert between mix proportions by mass and by volume (BSI, 1995).

Table 8.2 Limits on the mechanical properties of aggregates

Type of concrete	BS 882 requirements		EN 12620 category recommended for United Kingdom
	Ten per cent fines value (minimum) kN	Aggregate impact value (maximum) %	Los Angeles value (maximum) %
Heavy-duty concrete floor finishes	150	25	40
Pavement wearing surfaces	100	30	40
Others	50	45	40

Particle density is largely a consequence of geological type. Aggregate producers can have some influence, for example by processing to remove lightweight particles with a particle density of more than 2000 kg m^{-3} (ASTM, 1994). Particle density affects concrete yield and density.

Water absorption is usually determined at the same time as particle density. Absorption affects mix design and density, and is indirectly related to shrinkage, soundness, frost susceptibility and general durability (Smith and Collis, 1993). Absorption limits are rare, though may be useful if a link to an undesirable concrete property such as frost susceptibility is established, because absorption is easily determined.

Bulk density is determined by filling a container of known volume with aggregate. The mass of aggregate required to fill the container divided by the volume gives the bulk density. The value is usually determined as loose bulk density, that is, the aggregate is uncompacted. The British Standard describes and the European Standard allows the determination of bulk density in other, compacted conditions.

Particle density and absorption are determined by saturating the aggregate and weighing it in a saturated and surface dry condition and an oven-dried condition. Volume is determined from water displacement.

8.11 Drying shrinkage

The drying shrinkage of concrete is usually due to movement of the cement paste, because most natural aggregates do not undergo any appreciable drying shrinkage (BRE, 1991). Indeed, most aggregates restrain concrete shrinkage because they are less elastic than the cement paste to which they are bonded. Concretes with higher aggregate contents shrink substantially less than cement-rich mixes all else being equal.

Some rock types undergo unusually large volume changes on wetting and drying (BRE, 1991). These rocks are typically weathered, and contain clay or mica minerals. The absorption of the rock is sometimes high (Smith and Collis, 1993). If processed into aggregates, these rocks produce concrete with poor volume stability and a tendency to deflect and crack.

The British and European Standard tests for drying shrinkage of aggregates are similar. Concrete prisms are manufactured using the aggregates under test. The mix proportions of the concrete are fixed. The prisms are cured in water for five days and their length is measured. The prisms are then oven-dried and measured again. The difference between the two measurements expressed as a percentage of the final prism length is the drying shrinkage.

Aggregates giving a concrete drying shrinkage of 0.075 per cent or less are suitable for all concreting purposes. Aggregates with higher shrinkages are only suitable for concrete where complete drying out never occurs, for mass concrete surfaced with air-entrained concrete or for symmetrically and heavily reinforced members not exposed to the weather (BRE, 1991).

8.12 Soundness

A sound aggregate is able to resist stresses induced by environmental or climatic conditions. Some aggregates degrade or disintegrate under the action of cycles of freezing and thawing or salt weathering. Susceptible aggregates are usually microporous, that is, they have high absorption arising from a large proportion of fine pores (Neville, 1995; Sims and Brown, 1996). Freezing or salt crystallization can induce considerable stresses in fine pores that may lead to degradation. Unfortunately, there is no definite correlation between the soundness of aggregate and its performance in concrete. This is partly due, for example, to the resistance of concrete to cycles of freezing and thawing being affected by factors such as the degree of saturation of the concrete, its maturity, its quality and the severity of its exposure conditions as well as aggregate soundness.

British and European Standards contain the magnesium sulfate soundness test (Figure 8.6). In this test, the 14–10 mm fraction of the aggregate is placed in a wire basket and soaked in saturated magnesium sulfate solution. The sample is oven-dried to crystallize magnesium sulfate in the aggregate pores. The soaking and drying is repeated five times. The magnesium sulfate soundness value is defined in British Standards as the percentage of the sample retained on the 10 mm sieve after test. In European (and ASTM) Standards

Figure 8.6 Aggregate soaking in saturated magnesium sulfate solution during soundness testing.

the value is the percentage passing the 10 mm sieve after test. The procedure can be modified to examine other aggregate fractions, replacing the 10 mm sieve with the appropriate lower size sieve for the fraction.

A European Standard test subjecting aggregates to cycles of freezing and thawing is also available. There is no experience of this test in the United Kingdom.

8.13 Thermal properties

The thermal coefficient of expansion of aggregates typically ranges from 2 to 16 micro-strain/°C in normal temperature ranges (Harrison, 1992). The coefficient for silica minerals is higher than that for most other minerals. The expansion coefficient of an aggregate is therefore broadly related to its silica content, but wide variations can occur. Limestone is usually low in silica and has a low coefficient, so it may be specified for concrete when a particular risk of early-age thermal cracking has been identified.

The expansion coefficient of cement paste is higher than that of aggregate. Because concrete consists mainly of aggregate, the thermal coefficient of expansion of concrete is largely determined by the coefficient of the aggregate and the aggregate content. Limestone is the preferred aggregate for concrete when the expansion coefficient of the concrete needs to be kept low. While aggregate type is important, mix design, moisture condition and the age of the concrete are also significant influences. For these reasons, it is difficult to produce concrete with a thermal coefficient of expansion of less than 8–10 micro-strain/°C.

All aggregates produce fire-resistant concrete and the degree of fire resistance is related to aggregate type. Aggregates containing high levels of quartz offer poorer resistance, because quartz undergoes an expansive solid phase change at high temperature (Neville, 1995). Limestone contains little or no quartz and therefore offers better fire resistance. Lightweight aggregates are usually stable at high temperatures, have low coefficients of thermal expansion and provide heat insulation. They produce concrete with very good fire resistance (for more information see Chapter 5, Volume 2 in this series).

Thermal coefficient of expansion can be determined using strain gauges and a temperature-controlled cabinet. The change in the strain of the aggregate is measured as temperature changes and is used to calculate thermal coefficient of expansion (US Army, 1963).

8.14 Fines content

Fines are defined in British Standards as materials passing a 75 µm test sieve. European standards draw a distinction between sand and fines at 63 µm. This material may be very fine sand, silt, dust or clay. A moderate fines content fulfils a void-filling role, and aids cohesion and finishability. Excessive fines, especially fines consisting of clay, increase water demand and reduce the aggregate–cement paste bond. Both of these effects result in reduced strength.

Fines content is usually determined in conjunction with particle size distribution. The dried sample is washed over a guarded 75 µm (or 63 µm) test sieve (guarded with a nested

1.18 mm test sieve to prevent coarser particles damaging the 75 µm sieve) and then redried. The loss in mass is equivalent to the fines content.

Generally, more fines can be tolerated in crushed rock than in sand and gravel, because the fines consist of rock powder rather than clay or silt. BS 882 limits and recommended EN 12620 categories are shown in Table 8.3.

Table 8.3 Limits for fines content

Aggregate type	BS 882 limit	EN 12620 category recommended for United Kingdom
	(max % passing 75 µm)	(max % passing 63 µm)
Coarse gravel	2	1.5
Coarse crushed rock	4	4
Gravel sand	4	3
Crushed rock sand	16	16
Crushed rock sand in heavy duty floor finishes	9	10
All-in gravel	3	3
All-in crushed rock	11	11

8.15 Impurities

So far, this chapter has considered properties of the aggregate relating to the intrinsic properties of the mineral deposit and the characteristics of the processed product. Most aggregates also contain traces of materials that are not aggregate. These materials are called impurities and generally impart no benefit to concrete manufactured using the aggregate. Chlorides, sulfates and shell will be considered briefly. The effects of some other impurities are tabulated.

Chlorides occur in marine aggregates and some coastal and estuarine deposits. Processing removes the majority of chloride, but a residual amount usually remains in marine aggregate products. High levels of chloride can accelerate the set of fresh concrete and can lead to damp patches and efflorescence on hardened concrete as the salt migrates to the surface. However, the main problem with chlorides is that their presence increases the risk of corrosion of embedded metal. The risk of corrosion is related to the chloride ion content of the concrete expressed as a proportion of the cement content. It is not, therefore, possible to specify generally how much chloride can be tolerated in aggregate. But because aggregate constitutes such a large proportion of concrete, even relatively small percentages of chloride in the aggregate can contribute fairly large amounts of chloride ion to the concrete mix. For this reason, guideline limits on chloride content of aggregates tend to be low (BS, 1992). Guidelines are related to cement type and end use of the concrete because the risk and possible consequences of corrosion depend upon these factors. BS 882 guidelines are shown in Table 8.4.

Sulfates are found in natural ground, usually associated with ancient clays (Thaumasite Expert Group, 1999). On rare occasions these sulfates may be present as impurities in aggregates. High levels of sulfate can interfere with cement hydration in fresh concrete but the major concern is 'normal' sulfate attack of hardened concrete. (Other forms of sulfate attack are described in Volume 2, Chapter 12 in this series. In wet conditions, sulfates can react with components of Portland cement. The process is expansive and may

Table 8.4 Guidance on the chloride content of aggregates

Type and use of concrete	Maximum chloride ion content expressed as percentage by mass of combined aggregate
Prestressed concrete and heat-cured concrete containing embedded metal	0.01
Concrete containing embedded metal made with sulfate-resisting cement	0.03
Concrete containing embedded metal made with cements or combinations other than sulfate-resisting cement	0.05
Other concrete	No limit

lead to cracking and spalling of concrete. The risk of sulfate attack is related to the sulfate content of the concrete expressed as a proportion of the cement content. BS 882 requires producers to provide information on sulfate content of aggregates when required, to allow concrete producers to assess the amount of sulfate contributed to the mix by the aggregates. BS EN 12620 limits the sulfate content of natural aggregates to 0.2 per cent.

Shell is found to a greater or lesser extent in most marine aggregates. High levels of shell in aggregate can reduce the workability of fresh concrete and can make concrete more difficult to finish. The surface finish of hardened concrete may be affected by shell, and there is some evidence that hollow shells near the concrete surface may make concrete more susceptible to freeze–thaw damage (Lees, 1987). BS 882 limits the shell content of 10 mm aggregate to 20 per cent and coarser aggregates to 10 per cent. BS EN 12620 limits the shell content of all coarse aggregates to 10 per cent. Neither Standard limits the amount of shell in fine aggregate (see also Table 8.5).

Table 8.5 Some other impurities and their effects

Impurity	Effect on fresh concrete	Effect on hardened concrete
Acid soluble material in sand	None	Reduced skid resistance of pavement quality concrete
Alkali reactive silica	None	Risk of alkali–silica reaction
Swelling clays	Increased water demand	Reduced strength
Reactive iron pyrites	Possible reduced yield	Surface staining
Mica	Increased water demand	Reduced strength
Organic matter	Possible retardation	Possible reduced strength
Coal and lignite	Possible retardation	Surface staining, pop-outs
Soluble lead or zinc	Possible retardation	Possible reduced strength

8.16 Summary

Aggregate is the main constituent of concrete and the properties of aggregate influence the properties of concrete. Petrological classification alone is an inadequate guide to aggregate performance. Any determination of aggregate properties requires samples to be taken. Correct sampling, sample reduction and sample preparation are essential if test results are to be meaningful.

Consistency of particle size distribution is important to the properties of fresh concrete. There is a more minor influence on hardened concrete properties. Larger maximum particle size benefits lean concrete but the optimum size is around 20 mm for normal strength concrete. Equidimensional particle shapes are best. Shape and texture affect water demand and aggregate–cement paste bond. Aggregate must be strong enough to withstand mixing, handling and compaction without degradation. In practice, the aggregate strength usually exceeds concrete strength considerably. Bulk density can be used as a check on the consistency of aggregate grading and shape. Particle density and absorption must be known for mix design purposes.

Most aggregates restrain concrete shrinkage. A few aggregates undergo large volume changes on wetting and drying, and may produce concrete with poor volume stability. Some aggregates are less able to cope than others with stresses induced by environmental conditions. Susceptible aggregates may disintegrate after cycles of freezing and thawing.

The thermal coefficient of expansion of concrete is largely determined by the aggregate. All aggregates produce fire-resistant concrete though lightweight aggregates are usually best.

Moderate fines contents are useful for cohesion and finishability of concrete. Excessive fines contents increase water demand and may interfere with the aggregate–cement paste bond. Occasionally, impurities are associated with aggregates. At best, impurities impart no benefit to concrete. If present in sufficient quantities, their effects can be detrimental. Aggregate is essential to the economy, stability and durability of concrete.

References

American Society for Testing and Materials (1994) *C123 – Standard Test Method for Lightweight Pieces in Aggregate.* ASTM, West Conshohocken.

Ballmann, P., Collins, R., Delalande, G., Mishellany, A. and Sym, R. (1996) *Determination of the number of sampling increments necessary to make a bulk sample for tests for aggregates.* European Community Measurement and Testing Programme.

BRE Digest 357 (1991) *Shrinkage of natural aggregates in concrete.* Building Research Establishment: Watford.

British Standards Institution (1985) *BS 812 – Testing aggregates. Part 103 – Methods for determination of particle size distribution – Section 103.1 – Sieve tests.* BSI, London.

British Standards Institution (1989a) *BS 812 – Testing Aggregates. Part 102 – Methods for sampling* BSI, London.

British Standards Institution (1989b) *BS 812 – Testing Aggregates. Part 105 – Methods for determination of particle shape – Section 105.1 – Flakiness index.* BSI, London.

British Standards Institution (1992) *BS 882 – Specification for Aggregates from natural sources for concrete.* BSI, London.

British Standards Institution (1995) *BS 812 – Testing Aggregates. Part 2 – Methods of determination of density.* BSI, London.

British Standards Institution (2003) *PD 6682 – Aggregates – Part 1: Aggregates for concrete – Guidance on the use of BS EN 12620.* BSI, London.

Dewar, J D. (1983) *Computerized simulation of aggregate, mortar and concrete mixtures.* ERMCO Congress: London.

Eurochip (1996) Project report. Buckinghamshire Highways Laboratory Planning and Transportation Department.

European Committee for Standardization (CEN) (1996) *BS EN 932 – Tests for general properties of aggregates. Part 1. Methods of sampling.* BSI, London.

European Committee for Standardization (CEN) (1999) *BS EN 932 – Tests for general properties of aggregates. Part 2. Methods of reducing laboratory samples.* BSI, London.

European Committee for Standardization (CEN) (2002) *BS EN 12620 Aggregate for concrete*, BSI, London.

Harrison, T.A. (1992) *CIRIA Report 91: Early-age thermal crack control in concrete (revised edition).* Construction Industry Research and Information Association, London.

Highways Agency (1998) *Manual of Contract Documents for Highway Works.* Volume 1: Specification for Highway Works. The Stationery Office: London.

Lees, T.P. (1987) *C&CA Guide: Impurities in concreting aggregates.* Cement and Concrete Association, Slough.

Miller, E.W. (1978) *The effect of coarse aggregate grading on the properties of fresh and hardened concrete.* Institute of Concrete Technology. Advanced Concrete Technology Project, Crowthorne.

Neville, A.M. (1995) *Properties of Concrete* (4th edn). Longman, Harlow.

Ryle, R. (1988) Technical aspects of aggregates for concrete. *Quarry Management*, April.

Sims, I. and Brown, B.V. (1996) Concrete aggregates. In Hewlett, P.C. (ed.), *Lea's Chemistry of Cement and Concrete* (4th edn). Arnold, London, 903–1011.

Smith, M.R. and Collis, L. (1993) *Geological Society Engineering Geology Special Publication No 9.* Aggregates. Sand, gravel and crushed rock for construction purposes (2nd edn). The Geological Society, London.

Stanley, C.C. (1979) *Highlights in the History of Concrete.* Cement and Concrete Association, Slough.

Teychenné, D.C., Franklin, R.E., Erntroy, H.C., Nicholls, J.C. and Hobbs, D.W. (1988) *Design of normal concrete mixes.* Department of the Environment, London.

Thaumasite Expert Group (1999) *The thaumasite form of sulfate attack: Risks, diagnosis, remedial works and new construction.* Department of the Environment, Transport and the Regions, London.

United States Army (1963) *CRD–C 125–63, Method of test for coefficient of linear thermal expansion of coarse aggregate (strain-gage method).* US Army, Vicksburg, Miss.

Further reading

Geological Society Engineering Geology Special Publication No. 9, Aggregates, is a comprehensive guide to aggregates for all construction purposes.

Lea's Chemistry of Cement and Concrete (4th edn) contains a substantial chapter on concrete aggregates.

Appendix   Relevant standards for natural aggregates

British Standards

BS 812	Testing aggregates. A standard consisting of twenty parts, giving methods for sampling, examining and testing aggregates
BS 882	Specification for aggregates from natural sources for concrete
BS 1199/1200	Specification for building sands from natural sources

European Standards

EN 932	Tests for general properties of aggregates. In six parts, referring to sampling, petrography, equipment and precision.

BS EN 933   Tests for geometrical properties of aggregates. In ten parts, covering particle size distribution, shape, texture, shell and fines.

BS EN 1097   Tests for mechanical and physical properties of aggregates. In ten parts, covering strength, density, absorption, polishing and abrasion.

BS EN 1367   Tests for thermal and weathering properties of aggregates. In five parts, covering freeze–thaw, volume stability and heat.

BS EN 1744   Tests for chemical properties of aggregates. In four parts, covering chemical analysis and alkali–silica reaction.

BS EN 12620   Aggregates for concrete.

Limits for aggregate properties in BS EN 12620 are generally expressed as a range of values related to categories. This allows specifiers to choose limits suitable for the materials, climate, uses and practice in their region of Europe. A Published Document (BSI, 2003) provides guidance and recommends categories for use in the UK.

Index